Pack, Band, and Colony

The World of Social Animals

JUDITH and HERBERT KOHL

Pictures by MARGARET LA FARGE

Pack, Band, and Colony
Farrar, Straus + Giroux, Inc
19 Union Square West
New York, N.Y. 10003

Man is a social animal—only one of many species of animals that are—and there is much to be learned about ourselves from observing the social life of other animals. Wolf packs range over forty square miles of territory, and often more when they are hunting. Sifaka lemurs spend ninety percent of their time in trees fifty feet above the ground. Termites live in groups numbering hundreds of thousands, and it is almost impossible to distinguish one individual from another in their natural habitat. All three are social animals, dependent on others in their group for survival.

Judith and Herbert Kohl, in this fascinating look at non-human societies, go beyond the obvious problems of scientific observation to the heart of the matter. Wolves don't like to be observed and, as often as not, avoid or even observe their observers. Lemurs can be found only on a small island off the coast of Africa and appear to be a uniquely non-violent species. Termites live in total darkness, and so, when observed in the light of day, they behave abnormally. The Kohls expand on such observations as these to suggest the complexity of scientific examination, while going on to explain the value and, for the observer, the achievement possible in overcoming the problems.

This is a book for those who are fascinated by life around us and wonder at its relation to our own. Black-and-white detailed line drawings bring the animals described into sharp focus. Intertextural glossaries provide easy reference and clarification for the specialized concepts presented.

JUDITH and HERBERT KOHL won the National Book Award for children's literature for their book *View from the Oak*. MARGARET LA FARGE has illustrated several books, including *The Evolution of Culture in Animals* and *Penguins: Past & Present, Here & There.*

Black-and-white pictures throughout
Ages 12 up / 7 × 9 / 114 pages / ISBN 0-374-35694-7 / $10.95

Just Published

PACK, BAND, AND COLONY

PACK, BAND, AND COLONY

The World of Social Animals

JUDITH & HERBERT KOHL

Pictures by Margaret La Farge

Farrar · Straus · Giroux

NEW YORK

Published simultaneously in Canada by McGraw-Hill Ryerson Ltd., Toronto

Printed and bound in the United States of America by
the Murray Printing Company

Designed by Tere LoPrete

FIRST EDITION, 1983

Library of Congress Cataloging in Publication Data

Kohl, Judith.
Pack, band, and colony.
Summary: Discusses the study of social animals and three particular groups of them: wolves, lemurs, and termites.
1. Animal societies—Juvenile literature.
[1. Animal societies] I. Kohl, Herbert R. II. La Farge, Margaret, ill. III. Title.
QL775.K58 1983 591.52'4 82-20951
ISBN 0-374-35694-7

Special thanks to Professor Glenn D. Prestwich and Alison Richard for their help in supplying research materials

We would like to give special thanks to all those observers of animal life whose insights we have drawn on in writing this book. We would also like to thank our editor, Margaret Ferguson, and our typist, Lorna Cordy, for the time and energy they have put into this book.

—Judith and Herbert Kohl

CONTENTS

SETTING OUT

The ravens filled the morning air with the sound of their raucous caws

One day last spring Peter, a young friend who was visiting us for a few days, walked out to our chicken coop to get some eggs. But before he had time to get inside the coop, he heard a rooster crow high up in a nearby fir tree. He looked around to check on our only rooster—perhaps he had escaped? No, he was in the enclosure as usual, warily cocking his head. Within seconds there was another cock-a-doodle-doo, quickly followed by the *flap flap* of shiny black wings carrying a raven up and away over the top of the hill and out of sight.

Crows and ravens belong to the Corvid family. This American crow is smaller than a raven

That night at dinner Peter told us of his strange experience. It was the first time we had heard of a raven imitating a rooster, although we were aware that the raven's cousins, crows and magpies, can and do imitate many sounds. Assuming Peter had heard right, we speculated on what reasons a raven might have to imitate a rooster. Was it trying to attract the hens out of the coop so that their eggs would be exposed? (Ravens love to eat eggs.) Or was it attempting to entice a chick or even a hen away from the rest of the flock so that it would be unprotected? Or was it mocking the rooster just for the fun of it?

Yellow-billed magpies, like the one above, also belong to the Corvid family

Our speculations were interrupted when our son, Josh, excitedly insisted he had another, equally strange raven-rooster story to tell. He had seen our rooster give a piece of bread to a raven who was perched on a rail just outside the chicken-coop fence. We didn't know whether to believe that was what had actually happened, but had to admit that Josh watched the goings-on in the chicken coop regularly and frequently reported unusual occurrences that we later observed ourselves. Perhaps he really had seen the rooster pass some food to a raven.

Until we heard these two raven stories, we hadn't thought much about ravens, although we were certainly aware of their presence. They frequently fly over our house and garden, and it is even possible to hear the *whoosh* of their wings. Their dark, glossy feathers, their raspy clicks and caws, and their size make them an imposing, even ominous presence.

Ravens are about two feet long and their wingspan is over four feet

From then on we decided to watch ravens more closely. There are always lots of ravens at our county dump, so early one morning we parked our truck outside the closed gates, climbed over the fence, and as quietly as possible walked down the hill toward the dumping areas. The morning air was filled with the sound of raucous caws. As we approached, dozens of large black birds lifted off the rubbish heaps and angled up to the highest branches of the trees that ring the dump area, where they settled onto branches and joined the ravens already in the treetops. As long as they could see us, not one bird flew to the ground. They clicked, clucked, cawed, and rattled at one another from tree to tree. Occasionally two or three would rise from a treetop and soar off. They were probably tired of waiting for us to leave and had decided to feed somewhere else.

We tried to trick the ravens by walking back up the road and returning as quietly as possible to see if they would fly down to the rubbish heaps, where it was easier to observe them. When we returned ten minutes later, there wasn't one raven on the ground. They were still sputtering away far up in the trees, but now more and more of them began to fly toward the west and southwest. They flew in groups of three and four mostly, and by the time the dump opened fifteen minutes later, there were hardly any left. Except when we've purposely gone to look for ravens when the dump was closed, we've never seen them there again. There are, however, shiny black feathers on the bare ground in the dump almost all the time.

By this time we were obsessed with learning about ravens and ran out of the house whenever we heard or saw a raven outside. We began carrying binoculars in our car just in case we spotted a raven or two while driving, and discovered a stand of old cypress trees near the ocean where ravens congregated. It was beautiful to watch them circle high above the trees and the cow pastures that run down to the sand dunes. They hopped up and down in the pastures, and we learned from a book that they hop agitatedly when they're angry.

We began to read about ravens in bird books and in books on mythology, and looked for references to ravens in poems and novels, and for raven stories in newspapers, magazines, and technical journals.

We discovered that ravens have a mythological and historical past. People have been observing and thinking about them for thousands of years. For example, the raven was a prophetic bird for the ancient Greeks, the Arabs, and for many people in India, Western Europe, and Russia. The Arabs called the raven Father of Omens, and believed that if one sees a raven flying to the right, it means good luck, and if to the left, bad luck.

In Norse mythology Odin, the god who championed kings, warriors, and poets, had two pet ravens that sat on his shoulders. One was Huginn (Thought) and the other Munnin (Memory). These birds left Odin's shoulders each morning and flew around the world to watch and listen to everything that went on. Each night they would fly back to Odin's shoulders and whisper everything that they had seen and heard into his ears. These ravens were birds of war and provided information for Odin's kings and warriors to use and for his court poets to sing about.

The raven is also a central figure in the tales of the Eskimos and the Pacific Northwest Indians. In these stories a character called Raven created and helped people, although sometimes he tricked them as well. Some Pomo Indians, who live in Northern California, still tell an ancient raven creation

A Pomo myth tells how Father Coyote created people from a raven's feather

myth. According to this myth, the Pomo were created by Father Coyote, a trickster figure who made mounds of earth and placed a raven feather in each mound. Using his powers, Coyote breathed life into the mounds, which then became people. Thus, the spirit of the raven is in people.

Some people have perceived ravens as positive spirits representing the generosity of God. In the Bible, ravens feed Elijah and many saints: St. Paul the Hermit, St. Oswald, and St. Benedict are all pictured as being helped by ravens. One tale relates that St. Paul the hermit was fed a loaf of bread by a raven. Reading an account of that legend reminded us of Josh's story of our rooster passing food to a raven. All these myths and tales are probably based on actual observations. Throughout history people have watched animals, tried to learn from them, and acknowledged them as other living beings to share the world with. And often people did not separate their views of the world and their own needs from the conclusions they drew from their observations of animal life.

Our friend Peter is sure ravens have a reason to imitate roosters. Josh thinks our rooster has something going with a raven—whatever it may be. The local farmers despise ravens because they think they peck out the eyes of newborn lambs, and some Pomo respect ravens as part of their traditional heritage. We also want to learn about ravens, to study them, and to find out about their everyday lives, habits, and social groupings. Yet the ravens refuse to cooperate with us and confirm or deny any of our speculations, stories, or impressions. We can't get close enough to study them, possibly because there are many hunters around who have taught the ravens to be wary and suspicious of humans. Only once did one come near. That was on a local beach. Our guess is that human beach types are not gun types and that ravens know this.

After many futile attempts at approaching ravens and organizing our observations, we gave up trying to find out everything about the ravens who live around here. We have a small black notebook with entries like: "Three ravens circling high, look like they're playing follow the leader, leader drops way down, others follow, circle up again moving over toward hills (toward our house!), too far away to see any more." "Tried to get near roosting place on bicycles today. Ravens flew away and scattered. We got no closer than on foot." We have some fair tape recordings of the sounds they make. They caw when they fly, make a metallic rattling sound from treetops, and when they are in large groups roosting in trees, they make a great variety of guttural and short squawking noises that sound to us almost like human conversation. But we don't have any idea of what they are saying.

It was surprising to learn, even after all our reading, how little people actually know about the lives of ravens, and exciting to know that there are discoveries still to be made, from patient observations that should and most likely will be made at some time. Science so often seems to be something other people have already done, and yet here is a field where, with patience and, to borrow from the ravens, Thought and Memory, Peter, Josh, and any of us can add to what is already known and discover new things about the natural world.

A friend made a suggestion about furthering our knowledge of ravens. He suggested we steal a young raven from its nest and raise it in a cage. Then we could really get a close look. We tried to explain to him that we didn't want to get a close look at a raven, we wanted to get a close look at raven life. A lone raven is as sad as an ant isolated from its hill and put in a jar, or a person in solitary confinement. People, ants, ravens are social creatures. We know our own lives are distorted, saddened, and impoverished by isolation. To understand the lives of social animals you have to observe them in their natural worlds, not isolate them.

There are people who have devoted their lives to the study

of the social worlds of various animals, who have traveled thousands of miles and spent thousands of hours observing the lives of animals in their natural settings. This book is about some of these voyages, and about the animals themselves, and how they share the world with each other.

SOCIAL LIFE

The Individual and the Group

Frequently a male-female encounter between shrews begins with fighting

All animals have to be sociable at least once in their lives, or they wouldn't mate and leave offspring. Any species that suddenly became totally unsociable would quickly become extinct. At one end of the scale are animals that live solitary lives except when they mate. The female lays fertilized eggs in a suitable place and then leaves them to hatch unattended. Many insects fall into this category, but even some vertebrates, like turtles and many snakes, are extremely solitary. With any of these animals, the female has to deposit her eggs in a place that is close to an appropriate food source. The female Mediterranean fruit fly, for instance, lays her eggs in growing fruit, and it isn't until the eggs hatch and the larvae have begun to seriously damage the fruit by feeding on it from the inside that anyone is aware that they are there. Some insects have sharp ovipositors that pierce the bodies of other living insects so that their eggs are laid inside the host. When the larvae hatch, they spend most of their lives living off the host insect by eating it.

Species.
Individual animals of the same animal group that can interbreed with each other. They usually eat the same foods and have the same enemies.

Vertebrates.
Animals with backbones, including fishes, amphibians, reptiles, birds, and mammals.

Other animals have even more elaborate ways of making certain their young will have a food supply when they hatch. Digger wasps lay their eggs in holes and then kill insects to put in these holes. When the larvae hatch, they have an adequate supply of insects to feed on. There are several species of solitary bees that enter the nests of other bee species. They find their way to cells that already contain eggs and lay their own eggs in these cells. Some of these parasitic bees destroy the original eggs when they lay their own, while others leave the killing of the host larvae to the parasite larvae when they hatch. In either case, the parasite bee's offspring are assured of care and food. In this way these bees behave similarly to the European cuckoo bird, which lays eggs in other birds' nests. Sooner or later the young cuckoo makes sure it gets all its foster parents' attention by pushing its nestmates out of the nest with its shoulders.

A baby European cuckoo pushing an egg out of its nest

Another level of sociability is found among animals that live solitary lives except during mating but care for their young. Many mammals and fish live this way. Among fish who care for their young are the sticklebacks. The male stickleback builds a nest and entices a female into it. She lays her eggs and swims away. The father guards the eggs, and when they hatch, he brings food and protects the young fish from predators.

Moles and shrews, small insect-eating animals called insectivores, seem to want to have nothing to do with any other animals, even their own species. Whenever male shrews encounter each other, they squeak and bite. The loser falls on its back and shrieks, and the victor runs off. Sometimes the victor even kills and eats its opponent. Frequently a male-female encounter between moles or shrews begins with fighting and chasing, and it is some time before they overcome their normal aggressive behavior and can get along well enough to mate. Afterward they once again have nothing to do with each other. The female bears the young and raises them until they are mature enough to go their own solitary ways.

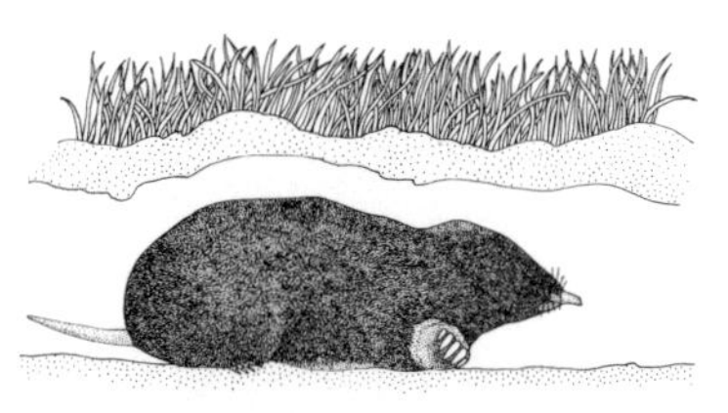

Moles have big forelimbs and conical noses for burrowing. Their eyes are barely the size of a pinhead, and they live mainly on earthworms

The next level of sociability is best characterized by a mated pair of adults which raise their young together. Many birds live this way. Some pair matings, such as sparrows', last for only one season of breeding and caring for the young; other pairs, like pheasants and Canada geese, usually mate for life.

Among some kinds of birds there are times of the year when the pairs break up and all the area's birds of that species form a large flock. Sometimes the adults remain paired but join the larger group, which is made up of other pairs and adolescent youngsters. For those months when the birds are freed from the demanding responsibility of feeding their young, they congregate together and range more freely—or even migrate in large groups to wintering spots often thousands of miles from their home territories.

The animals we will focus on in this book are more sociable than any we have discussed so far. Wolves, lemurs, and termites—our prime subjects—are very different from one another in many obvious ways, but they all live in groups that cooperate to raise their young and defend the group from outside enemies. Wolves and lemurs, like humans, do not have an absolute need to live in a group after they mature. Some individuals leave their group and exist as loners, but living in close association with other members of the same species is the norm. Termites and other social insects, however, cannot survive as individuals for any length of time.

There are many different kinds of social groups. A wolf pack is different in structure from a lemur band or a termite colony. There are even many different kinds of wolf packs, just as there are a variety of ways humans live together. The nature of the life of an animal species—its food, its native intelligence, its predators, whether it lays eggs or gives birth to live young, its body size—is going to influence the way a group of animals live together.

We must also know something about the physical capacities and limitations of an animal's body and senses before any interpretation of its social behavior can be made. Some animals can hear sounds we cannot; others have eyes but are blind in the light. Some feel the world around them with their body hairs, while others with hard outer shells have little sensation in parts of their bodies. Social relations, which involve contact between individuals, are determined by perception and physical capacities. This fact may seem trivial, and yet it can lead to discoveries that astonish people, such as that moths can communicate over miles by scent and that dogs and cats can sometimes find their way home over long stretches of unfamiliar territory.

To get a sense of the range of possible social relationships it might be useful to play with a simple physical model that illustrates forms of dependence and independence. Imagine that you have five cube-shaped building blocks. How many different ways is it possible to arrange them? There is probably no limit to the number of possible arrangements, especially if very small changes in position are allowed. Some arrangements, however, are more interesting than others, since they change the role each block plays with respect to the other four blocks. Consider a very simple arrangement. One block is placed in each corner of a table, and the fifth block is placed in the middle, creating this pattern:

Dependence.
A relationship that exists between several objects or animals if changes in one of them produce changes in the others. If changes in one produce no changes in the others, then the objects or animals are said to be *independent* of each other.

The five blocks are not touching. You can remove any block without moving the other blocks at all. This is true of all arrangements that involve no block touching any other block.

Now, move the blocks together and create a row of five blocks like this:

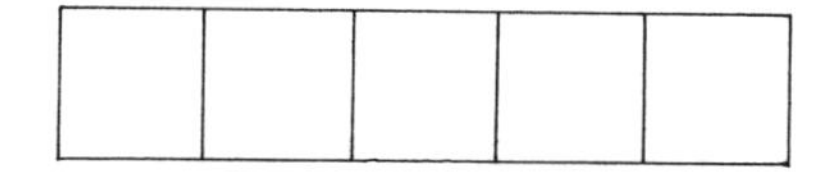

Try to remove the end blocks without moving any of the other blocks. If you're careful you can do it. Try to remove the middle block without affecting the others. It's possible but not easy. With modification in the arrangement, moving the middle block becomes much harder, since the blocks are much more closely related to one another:

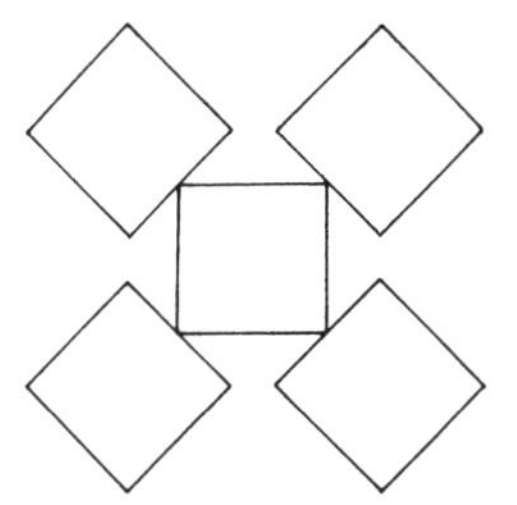

Now build a simple tower and try to remove the middle block without moving the others:

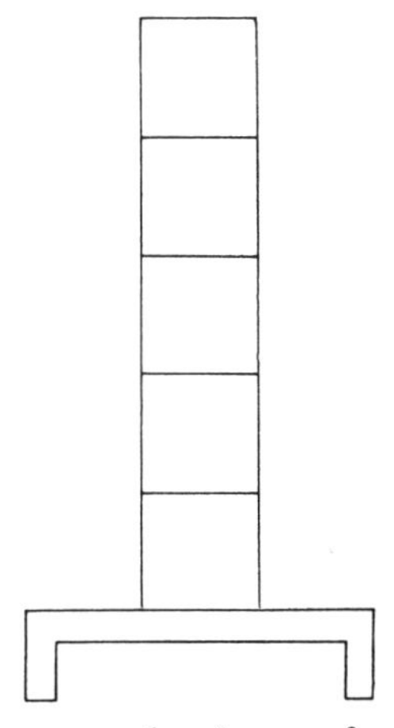

It's impossible so long as the law of gravity still operates. Removing the middle block leaves the top two with no support and nowhere to go but down.

Now imagine that instead of experimenting with blocks you are observing living creatures and trying to find out how their worlds fit together. You can look for relationships between individuals and see how they vary from group to group. For example, are all the individuals in a group separated from one another like blocks spread out on a tabletop, so that a change in the life of one doesn't have any effect on the others? Or is it a group where the individuals are so closely related that small changes can cause it to collapse like a tower made out of blocks?

Monkey fear

Human joy

For people as social animals, knowing how to discover relationships within a group is an important part of life. Whenever we meet a new group of people, we have to figure how they relate to one another and how they expect us to relate to them, as well as what kind of behavior indicates such important attitudes as friendship, hostility, respect, and deference. This is true at parties, in school, at work, with one's neighbor. Often the discovery of group relationships functions automatically and we take it for granted. For example, when we enter a new school or begin a new job, we automatically "check it out" without thinking much about the techniques we're using. However, when we observe groups of animals to find clues about how they relate to one another, we have to be more conscious of what is happening. We have to learn the "vocabulary" they use among themselves and be careful not to confuse their behavior with our

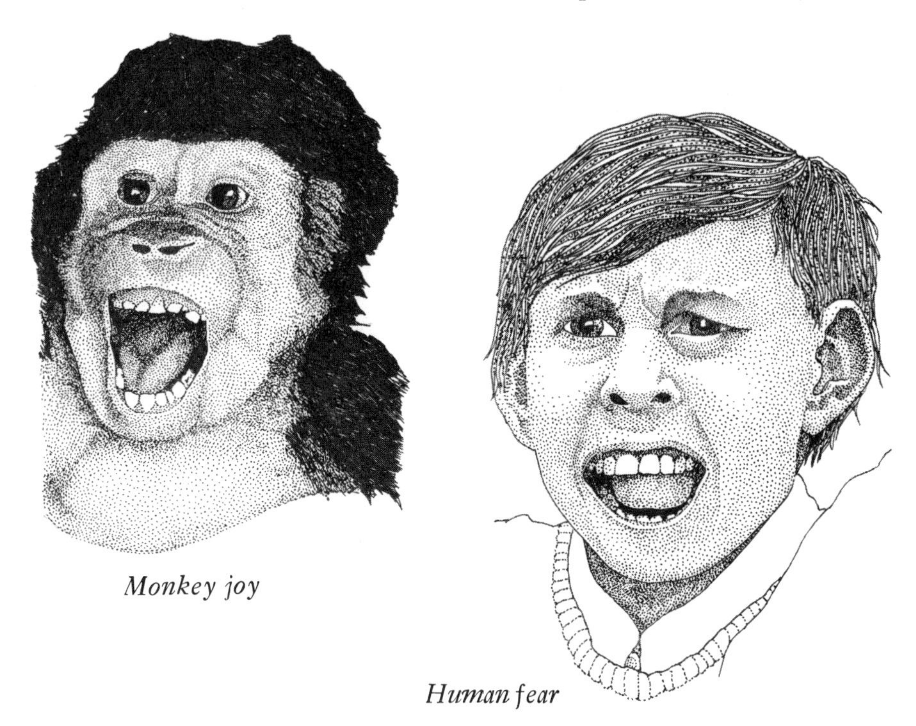

Monkey joy

Human fear

own. For instance, a monkey "grin" indicates fear, not pleasure, while a wide-open mouth that shows bared teeth can be a sign of joy rather than of threatening behavior.

Humans and animals express their feelings toward other members of their species, and in some cases unrelated species, in a variety of ways. Visual signs are easy to observe—a friendly wave of the hand from a friend, the stiff arching of a cat's body, the slow waving of an insect's antennae. When properly understood, the sounds we and other animals make tell us about how those animals are feeling or how they express needs and feelings toward each other. The roar of a crowd cheering a football team or a flock of blue jays screeching at a cat gives us information about what is happening within a group. Smells and scents can indicate specific attitudes and ways of relating. Ants mark food trails with odorous liquids so that other ants can follow the scents to a food source. Dogs give information about themselves to other dogs by leaving a trail of urine odors. Although humans do not depend on scent as much as other animals do, we use perfume, enjoy the smell of good food, and are concerned about the way our bodies do or don't smell.

Perhaps the most subtle kind of expression, and the one most difficult to observe and interpret, is that which begins inside the body and is not under our or other animals' control —blushing, body and facial tics and twitches, stomach rumblings, hair raising or diarrhea caused by fear. Although these internal expressions are involuntary, they can give us clues about how animals feel and respond to their world.

When observing social animals, it is not enough to observe a single individual. One has to be aware of the group, watch for meetings, greetings, fights, flirtations, and other forms of social interaction. It is easy to get a completely wrong impression of individual and group behavior by focusing on a single individual or especially on a small sample of the individual's behavior. We learned that painfully several years ago. We used to have a dog named Sandy which was temperamentally incapable of harming a baby, much less an adult. All he wanted to do was retrieve sticks and be around people. He didn't even seem to be very interested in food. When we lived in a city, that was just about all of Sandy's life we perceived. When we moved to the country Sandy didn't seem to change much. However, he occasionally disappeared for several days at a time, usually returning home early in the morning looking tired and slightly deranged. We assumed he was chasing female dogs. But then one morning Sandy left a deer skull in front of our house, and several weeks later the flank of a freshly killed deer was on our doorstep. It was clear that he was bringing home his share of a kill.

A few weeks later a local sheep farmer dropped by and told us that a pack of dogs had killed a half dozen of his sheep and that he couldn't afford to have it happen again. He knew Sandy was gentle at home, a nice obedient dog. But, he continued, put a golden retriever together with two Labradors, a pit bull, and a terrier, and you have a pack of killer dogs. Fortunately, we could prove that Sandy had been home for the last week. But we also knew the farmer was right and that even if Sandy was innocent this time, we had to control him. We learned the signs he made when he was ready to join his friends in a hunting expedition: ears constantly erect listening for something we couldn't hear, restless pacing through the house, uncharacteristic whining. For the sake of his survival, as well as that of the sheep and deer, we had to chain him up during those times.

We have to admit we were curious about this wild social part of Sandy, and though we fantasized about following him with the pack and observing them, we knew that the first time they plunged into the woods we would be left hopelessly behind, limited by our two feet, erect posture, and lack of speed and stamina. We cannot run with dog packs, just as we cannot fly with ravens or burrow with moles. We cannot live inside anthills or beehives, cannot move in the treetops like monkeys. We cannot detect chemical trails like ants or smell sticks like dogs or spot prey like eagles. Our senses, our size and mobility limit what we can learn of the social life of animals.

Yet all these limitations provide challenges. How much can we learn? How can we extend our senses, increase our mobility, and strengthen our powers of observation using our intelligence and ingenuity? After all, binoculars, magnifying glasses, cameras, and microscopes extend the power of our eyes; tape recorders and sophisticated microphones extend the power of our ears; airplanes and land and sea vehicles extend our mobility; chemical analysis extends our ability to detect scent and chemical signals. We are limited in what we can learn about the social lives of other creatures, but the exact limits have not been established.

We can put microphones inside anthills, and someday we may even be able to put miniature video cameras inside a beehive. We can attach small transmitters to birds and wolves, and track them without harming them. We don't know what limits we will ultimately encounter in studying other creatures, although it is hard to imagine that we will ever be able to communicate with animals in their own languages and learn about their worlds directly from them.

Studying the social life of animals requires preparation as well as patience. What kinds of things do we want to look for? What equipment should we take on a voyage into the world of an ant or a bee, and what equipment for an exploration of the world of flying squirrels or sea otters? Each voyage has its own scale, its particular characteristics, its limitations. We have to prepare and equip ourselves to study animal life in as serious a manner as explorers like Rasmussen and Byrd did to explore the Antarctic, and as astronauts do now. We need to learn as much as possible about what is already known about the animals we want to study and the world they live in. We have to prepare ourselves physically, train our powers of observation and concentration, learn to choose and use the right equipment, and be able to keep accurate notes of our observations, as well as sketch and photograph what we see.

We also have to respect and learn from people who live in the constant presence of the animals we study. Often more knowledge of animal life can be acquired from people who know animals in an everyday way than from books and scientific journals. If eagle feathers are essential to a people's religious ceremonies, or ostrich eggs or wild boar the mainstay of a people's diet, it is necessary for these people to know an enormous amount about the behavior of these animals in order to hunt them down.

Finally, like any explorer, the explorer of animal life must expect the serendipitous. Serendipity means the unexpected, the unplanned, the unusual, an initially unexplainable encounter. The word was coined by Horace Walpole in his fairy tale *The Three Princes of Serendip*, the heroes of which "were always making discoveries, by accidents and sagacity, of things they were not in quest of." Serendipity, the ability to learn from the unexpected occurrence, is essential to exploration. What makes such voyages into the world of animals exciting is what isn't known about each animal, not what we already know. And since social life in the animal world is little known or understood, it is a wonderful field for scientific adventurers.

WHAT ANIMALS DO YOU LIKE?

The ingenuity of Lorenz's work was possible because he cared about his geese

If you plan to study animals over a length of time, it's important to have a personal affection for the kind of animals you observe and a delight in their natural environment. If you hate the cold and are terrified of bears, it won't make much sense for you to study polar bears in the Arctic. If you hate crawling on the ground and getting your clothes dirty, you shouldn't be studying swamp snakes or mosquitoes. The study of animal life has to originate with some personal feeling for a form of life. We love ravens and whales, admire aphids, identify with the wildness of coyotes and the ingenuity of spiders. It is our feeling for the animals that leads us to study them in the first place, but when we study them we must be careful to prevent these feelings from influencing our interpretations of our observations or our reading of research on their lives.

Many people who are free to choose the kind of work they do base their choices on love of something outside themselves, whether it be machines, animals, numbers, people, books, color, music, or movement. As Konrad Lorenz, who won the Nobel Prize in biology for his work studying the life of the greylag geese, said about his book of photographs of the geese, "This is not a scientific book. It would be true to say that it grew out of the pleasure I take in my observations of living animals, but that is nothing unusual, since all my academic works have also originated in the same pleasure." The patience and ingenuity involved in Lorenz's work were possible because he cared about his geese, perhaps even identified with them and wished he could fly the migration routes with their extraordinary flocks. But his affection for the geese, which is so clear in the pictures in the book, didn't cloud or distort Lorenz's scientific work. What it did was provide it with energy and personal satisfaction that allowed him to work for fifty years on a subject few would have thought of studying.

Not every student of animal life makes the right choice of animal. Sometimes the match between scientist and animal is so bad that it distorts the behavior of both. Nicholas Thompson, who studies the lives of crows, started out studying monkey behavior in a laboratory setting. He described his short career with monkeys in the following way:

> I was, by coincidence of geography and age, one of the first graduate students to specifically train himself for a career studying the social behavior of primates. I believed deeply in field study, but because of an irrational fear of leeches and lions, I set about doing a laboratory study in partial satisfaction of the requirements for a Ph.D. in Comparative Psychology. For three years, off and on, I and my monkeys performed together in the beautifully appointed, windowless basement laboratories of Tolman Hall at the University of California at Berkeley. My days were spent from nine to five testing and filming monkeys, while my evenings were devoted to coding and organizing the hundreds of thousands of frames of timelapse film which my labors were generating.
>
> From the horror of those days, I remember almost nothing about simian social organization. I remember the constant battle to keep man, monkey, and environment in their separate places. I remember the paraphernalia of sanitation: the stainless steel buckets, the steam hoses which behaved like sulky snakes when one tried to coil them, and the anti-T.B. foot tray which exuded

a pleasant odor of iodine but which left one's foot faintly wet for hours. I remember the virus B. anxiety, helped along by occasional ambiguous reports of deaths at other laboratories, deaths from monkey bite, or monkey scratches, or monkey spit, or perhaps even just monkey breath. Each of these hazards required a piece of garb to protect the investigator, so that the official standard outfit in our laboratory consisted of boots, coveralls, rubber gloves, paper caps, face masks, and goggles. Each worker in the laboratory omitted some part of this armor, based on a personal calculation which included as variables how much he hated the particular item and how recently he had read of a Monkey B. death. The goggles were a favorite omission because they fogged on the inside from one's breath and on the outside from the steam hoses. Those of us who omitted this particular item developed a covert, subconscious squint, which we deployed in the presence of monkeys to guard us against a lethal dose of monkey spit. Thus did we who had pledged our professional lives to understanding monkeys spend the majority of our energies in isolating ourselves from monkeys.

What is significant about this period is that I came to hate monkeys . . . these were incorrigible prisoners, and I was both warden and sole guard. All the while I worked with my subjects, I was promising myself never again to look at a confined monkey; since leeches and lions still held their terrible power over me, I had to recognize that I was promising myself never again to look at a monkey. Readers may conclude at this point that my reasons for leaving primate social behavior were too personal to be of general relevance. But personal reasons have a way of gathering logic, and many a scientist has been forced to make observations of catholic importance because of a personal bind.

There are probably some people who don't fear lions and leeches, who even like the feel of confronting monkeys in a windowless laboratory. People choose their animals to study, and their reasons are neither objective nor scientific. Edward O. Wilson, who is one of the best-known specialists on insect societies in the world, just happens to like bugs. In his serious and somewhat technical book entitled *The Insect Societies*, his affection and obsession occasionally emerge. In the section entitled "Symbioses with Other Arthropods," Wilson describes aphids that emit drops of honeydew which ants feed on. The drops are emitted when the ants stroke the smaller aphids with their antennae. In order to confirm the general theory about how ants get honeydew from aphids, Wilson put himself in the place of an ant. He says, of getting aphids to emit honeydew, "I have done this successfully myself with giant myrmecophilous coccids in New Guinea using one of my own hairs, and then collected the material directly from the insects and tasted it." Now imagine this famous scientist, a professor at Harvard University, pulling out a strand of hair and stroking a tiny aphid, then collecting its honeydew and tasting it. That, one might say, is a combination of love and science, a wonderful match of pleasure and technical work.

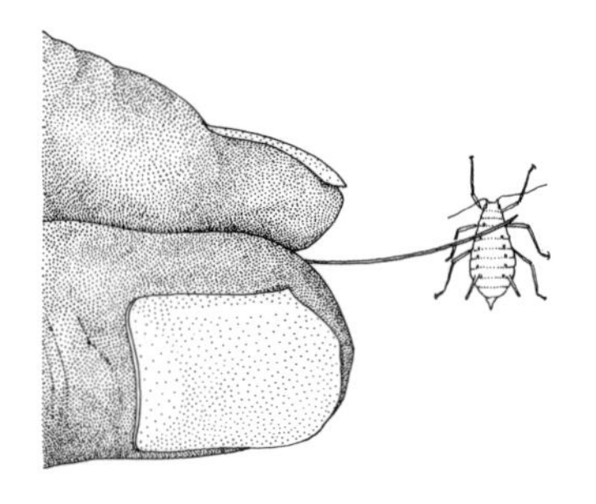

Tickling an aphid to get honeydew

Here's another example of Wilson's relation to his work. Later in his book, in a discussion about competition among different species of ants, he talks about how you can put down a wet cube of sugar and watch different ant groups struggle over the food. He says, "As a matter of fact, I often do this experiment as a form of amusement. On the streets of San Juan, Puerto Rico, to take one of many examples, anywhere from one to six species are attracted to the same sugar bait."

Wilson likes insects, Thompson prefers crows to monkeys, Lorenz cares about geese, and other scientists prefer bees or apes or people living in crowded cities. The choice of subject is not arbitrary. It comes from caring about the life you are studying. In this book we have chosen three social animals we care about to consider in detail—wolves, lemurs, and termites.

Wolves have always had a strong place in human imagination. From the frightening werewolf to the loving wolf that was supposed to have nurtured Remus and Romulus, the twin babies who, as adults, founded the city of Rome, wolves have always been portrayed as animals of strength, emotion, and intelligence. We wanted to know about wolves and about the people who have studied them, about the truth of the tales of killer wolf packs and lone wolves.

Lemurs interested us for quite different reasons. They exist only in Madagascar and the Comoro Islands, might soon become extinct, and are complex animals that evolved with no contact with monkeys, apes, or any other primates. They look to be part dog, part bear, part raccoon, part mouse, and part monkey. They are fascinating because they are so different, because so little is known about them, and also because some of their groups seem to have developed ways of avoiding violence and functioning peaceably and cooperatively.

Primate.
The order of mammals which includes all lemurs, monkeys, apes, and man.

We were compelled to learn more about termites because we once lived in a termite-infested house and needed to learn something about their habits. We feared and hated these insects that were eating our house's foundations until we discovered that there are termite species that build huge mounds that are homes for millions of termites and that these colonies could survive for decades. Reluctantly we came to admire them.

In the next three chapters we will take you on three voyages of discovery—one to the territory of a wolf pack, another to the jungles of Madagascar, where sifaka lemurs live, a third to termite cities in East Africa. Each voyage will require different preparation. Wolves range over a wide territory, sifaka lemurs spend 90 percent of their time in trees fifty feet above the ground, and termites are small insects who spend most of their lives within their nests. These conditions of life require different modes of observation and different equipment.

We hope each of these voyages will give you a sense, not just of the lives of these animals, but of what it is like to study them in their own environments.

THE WOLF'S WORLD

A wolf pack may range over 100 square miles

We have all grown up with myths, fairy tales, and stories about wolves. Almost all of these are about a dangerous animal who lurks around towns and preys on pigs, children, grandmothers, and defenseless sheep. Hardly any are about the lives that wolves really live. The idea of a wolf pack has been used by people to describe certain ways *we* behave; it has become an image for the cruel and predatory way some humans act when they pack together to do socially destructive things, as when teenagers act according to the rules of a violent gang or soldiers raid and destroy civilian villages. From the stories of hunters and pioneers we are also led to believe that wolves swarmed over the countryside on constant rampages, attacking everything in their path—buffalo, cattle, people. They are thought to be such a threat to human life and domesticated animals that in most parts of the world where the wolf used to live people shot, trapped, and poisoned them so extensively that there are now only a few places left where wolves exist in the wild. While they used to live throughout most of the Northern Hemisphere, they are now nearly extinct in Europe, and in the Americas are found only in Canada, Montana, Alaska, southeastern Texas, Minnesota, and the Upper Peninsula of Michigan. Occasionally lone wolves wander into the Western border states. Those are often young wolves moving out from established packs in British Columbia, Alberta, and Saskatchewan in Canada.

Except for the red wolf, which still survives in southeastern Texas, the wolves that live in North America are usually either timber or tundra wolves, depending on whether they live in or near forests or on the tundra of northern Canada and Alaska. The timber wolf is also sometimes known as the gray wolf. Their closest relatives are domestic dogs, dingos, coyotes, and jackals and more distantly foxes and African wild dogs. They are so close to domesticated dogs and coyotes that they can and do interbreed with them.

Sixty million years ago the first rodentlike mammals had probably already given rise to some kind of carnivore that lived in forests and hunted by chasing its prey. By 20 million years ago this group had divided into two distinct carnivorous groups—dogs and cats—and there were animals that could be recognized as ancestors of wolves. One million years ago the wolf's immediate ancestor made its appearance. It had a larger brain and a longer nose than its predecessor and may have had a primitive social structure and some cooperative hunting techniques.

Carnivore.
An animal whose diet consists primarily of meat.

Until about forty years ago most of our information about the habits and behavior of wolves came from men who made their living hunting and killing them. Since then another kind of hunter has appeared. These hunters were trained as biologists and zoologists. Some of them were hired by government agencies concerned with livestock protection to find out about wolves, so that more effective ways of killing them could be developed and put into effect. Others, out of scientific curiosity, have simply been trying to find out about how wolves live with one another as they hunt, eat, play, mate, and raise their young.

In this chapter you will read about several of these studies and the different ways scientists who made the investigations chose to pursue a difficult subject. You will also find out some of the things that people who live near wolves know about them. We will also create a picture of a wolf pack based on what these patient (and sometimes not so patient) observers have found out about the way wolves live together.

Learning about wolf packs is not an easy thing to do. The problem begins with the enormous size of wolf territories. A wolf pack may consider a 40-square-mile territory home and regularly range over 100 square miles. Another problem is the wolves' intelligence. Wolves used to allow humans to observe them from a distance, but as man with his guns and airplanes moved into remote territories, they have become increasingly cautious and secretive, although they do seem to know the difference between hunters and people who do not intend to hurt them. A wolf pack on Isle Royale in Lake Superior did not appear to be bothered much by the scientists who tracked their activities by following them in airplanes. As soon as some hunters began to shoot at them from airplanes, though, they suddenly became frightened of all airplanes, and the biologists who had been tracking them for years had a hard time finding any wolves in the open until a ban against aerial hunting was enforced. Soon after the ban the wolves once again trusted the biologists' planes and became less elusive.

One student who managed to make some interesting observations of wolf life is a Canadian named Farley Mowat. In the 1950s Mowat was sent by the Canadian government to take a census of wolves in the Northwest Territories and to collect information about their habits, diet, and effect on the caribou population. He described his life in the wilds of northern Canada in his book *Never Cry Wolf*. Mowat was flown into what seemed to him at first a barren wasteland where he was to do his surveys. He was equipped for the expedition with an absurd amount of government-supplied equipment, including camp stoves and tents, dog harnesses for nonexistent dogs, skis, and a canoe. Once he had arrived at his destination and discarded unnecessary equipment, it took him only a few days to discover a high rocky hill where he was certain there was a wolf den and a small pack of wolves. He figured that all he had to do then was find a spot to watch from. He found a small mound of rocks about 400 feet from the den where he could set up his telescope and still be hidden from the wolves' view.

Mowat then proceeded to look for the wolves through his telescope but could find absolutely nothing—no wolves, no movement of any kind. He was thoroughly discouraged. He describes it in his book:

> By 2:00 p.m. I had given up hope. There seemed no further point in concealment, so I got stiffly to my feet and prepared to relieve myself.
>
> Now it is a remarkable fact that a man, even though he may be alone in a small boat in mid-ocean, or isolated in the midst of the trackless forest, finds that the very process of unbuttoning causes him to become peculiarly sensitive to the possibility that he may be under observation. At this critical juncture none but the most self-assured of men, no matter how certain he may be of his privacy, can refrain from casting a surreptitious glance around to reassure himself that he really is alone.
>
> To say I was chagrined to discover I was *not* alone would be an understatement; for sitting directly behind me, not twenty yards away, were the missing wolves. They appeared to be quite relaxed and comfortable, as if they had been sitting there behind my back for hours. The big male seemed a trifle bored; but the female's gaze was fixed on me with what I took to be an expression of unabashed and even prurient curiosity. . . .
>
> "Shoo!" I screamed at them. "What the hell do you think you're at, you . . . you . . . peeping Toms! Go away, for heaven's sake!"
>
> The wolves were startled. They sprang to their feet, glanced at each other with a wild surmise, and then trotted off, passed down a draw, and disappeared in the direction of the esker. They did not once look back. . . .
>
> My thoughts that evening were confused. True, my prayer had been answered, and the wolves had certainly

> co-operated by reappearing; but on the other hand I was becoming prey to a small but nagging doubt as to just *who* was watching *whom*. I felt that I, because of my specific superiority as a member of *Homo sapiens*, together with my intensive technical training, was entitled to pride of place. The sneaking suspicion that, in point of fact, *I* was the one who was under observation had an unsettling effect upon my ego.

Mowat decided to set up a tent so that he could observe the wolves day and night.

> To begin with I set up a den of my own as near to the wolves as I could conveniently get without disturbing the even tenor of their lives too much. After all, I *was* a stranger, and an unwolflike one, so I did not feel I should go too far too fast. . . .
>
> I took a tiny tent and set it up on the shore of the bay immediately opposite to the den esker. I kept my camping gear to the barest minimum—a small primus stove, a stew pot, a teakettle, and a sleeping bag were the essentials. I took no weapons of any kind, although there were times when I regretted this omission, even if only fleetingly. The big telescope was set up in the mouth of the tent in such a way that I could observe the den by day or night without even getting out of my sleeping bag.
>
> During the first few days of my sojourn with the wolves I stayed inside the tent except for brief and necessary visits to the out-of-doors which I always undertook when the wolves were not in sight. The point of this personal concealment was to allow the animals to get used to the tent and to accept it as only another bump on a very bumpy piece of terrain. Later, when the mosquito population reached full flowering, I stayed in the tent practically all of the time unless there was a

strong wind blowing, for the most bloodthirsty beasts in the Arctic are not wolves, but the insatiable mosquitoes.

My precautions against disturbing the wolves were superfluous. It had required a week for me to get their measure, but they must have taken mine at our first meeting; and, while there was nothing overtly disdainful in their evident assessment of me, they managed to ignore my presence, and indeed my very existence, with a thoroughness which was somehow disconcerting.

Quite by accident I had pitched my tent within ten yards of one of the major paths used by the wolves when they were going to, or coming from, their hunting grounds to the westward; and only a few hours after I had taken up residence one of the wolves came back from a trip and discovered me and my tent. He was at the end of a hard night's work and was clearly tired and anxious to go home to bed. He came over a small rise fifty yards from me with his head down, his eyes half-closed, and a preoccupied air about him. Far from being the preternaturally alert and suspicious beast of fiction, this wolf was within fifteen yards of me, and might have gone right past the tent without seeing it at all, had I not banged my elbow against the teakettle, making a resounding clank. The wolf's head came up and his eyes opened wide, but he did not stop or falter in his pace. One brief, sidelong glance was all he vouchsafed to me as he continued on his way.

It was true that I wanted to be inconspicuous, but I felt uncomfortable at being so totally ignored. Nevertheless, during the two weeks which followed, one or more wolves used the track past my tent almost every night—and never . . . did they evince the slightest interest in me.

For several weeks Mowat continued his wolf watch, spying on the pack as it played and slept around its den and watching the steady coming and going of the adult wolves. As the summer wore on and the pups grew old enough to travel, the pack moved away and Mowat took to his canoe, for he had to get on with his assignment of estimating the size of the entire wolf population in the territory and collecting data on more than just one small wolf pack.

Mowat's patient observations enabled him to gather valuable information. He watched wolves eat, play, and rest around their den, and saw the patterns of their hunting expeditions. However, there was no way that Mowat or any other person could follow a pack of wolves on a hunt that ranged over forty miles on foot or skis. To get a sense of the range of wolves, imagine running twenty to forty miles through deep snow in search of food and doing it for eight to ten hours almost every day.

About the time Mowat was making his observations, other natural scientists were developing ways to track wolves' movements using radio transmitters. David Mech has spent twenty years following wolves in airplanes and pickup trucks. He frequently keeps track of them by first capturing them and putting radio collars on them, and then releasing them. Each collar sends out a particular radio signal so that each wolf's movements can be followed. Over a long period of time you can get an interesting picture of where wolves travel and how often, of how many travel, hunt, and live together, and for how long and during which seasons. Much of our information about wolves, particularly about their hunting habits, has been collected this way.

Radio collars are used to track wolves

Another source of wolf information has been wolves kept in captivity. Hardly anyone considers such information reliable, because almost everyone agrees that wolves in captivity do not act at all as they do in the wild. Even most people who raised wolf cubs, treated them kindly, and had deep affectionate bonds with them came to regret the lives the adult wolves spent when they had to be caged. Wolves need a large territory to range over, and they need the companionship of other wolves. Too many stresses build up among caged wolves, and since they are unable to run away from each other as they would in the wild, two angry animals can begin a fight that ends in the death of one of them. This is not to say no one has learned anything from captive wolves. Some fine observations of the early life of wolf cubs, for instance, are no doubt quite accurate as far as they go. The danger is in assuming that the *social* behavior of captive wolves is the same as it is among wolves in the wild.

Wolves in the wild occasionally kill other wolves—seriously injured or ill pack members are sometimes killed by the other members of the pack, and during territorial fights between packs an intruder may get killed by wolves defending their territory or den. Usually, though, wolves know when to quit a fight, and except for the sick or injured wolves that are killed, there is little if any fatal fighting among the members of a pack.

One of the most interesting sources of information about wolves is the people who live and hunt in parts of the world where wolves still live. In Alaska the Nunamiut Eskimo have always watched wolves. Robert Stephenson, a wildlife biologist, spent three years during the early 1970s living with the Nunamiut while he studied wolves and foxes. He soon realized that what he and other biologists were trying to find out about wolves was quite different from what the Nunamiut knew. The biologists would analyze stomach contents of dead wolves to see what they had eaten and count the scars on the uteruses of dead females to find out how many pups they had given birth to. The Nunamiut found this interesting but unimportant. Instead, they watched individual wolves and wolf packs as both people and wolves went about their daily lives. There was no generalized idea of "wolf behavior," but rather knowledge of how each individual wolf looked and behaved. Being hunters themselves, the Nunamiut knew how to learn from observing very small details: variations in footprints, shedded fur, scent markings.

A Nunamiut man was once asked if an old man or an old

wolf knew more about the mountains and foothills of the Brooks Range, where they both lived. He answered that Amaguk, which is what the Nunamiut call a wolf,

> is like Nunamiut. He doesn't hunt when the weather is bad. He likes to play. He works hard to get food for his family. His hair starts to get white when he is old. Young wolves, just like Nunamiut, run around in shallow melt ponds scaring the ducks.
>
> And Amaguk is tough, living at fifty below zero, through blizzards, for months without caribou. Like Nunamiut. Maybe tougher. And Amaguk is smart. He sets up ambushes for caribou. He sleeps high up on the ridges when there are humans around. He brings his pups to a kill but won't let them stay there alone. . . .
>
> Amaguk and Nunamiut like caribou meat, know the good places for caribou hunting. Where ground squirrels are good. Where to get raspberries. A good place for getting away from mosquitoes. Where lupine blooms first in May. Where that big rock is that looks like achlack, the grizzly bear. Where the creeks are still running in August. . . .
>
> The same. They know the same.

People learn about wolves in many different ways, and yet no single method provides a full picture of wolf life. Radio tracking, on-the-spot observation, physical analysis, and laboratory studies all have built-in limitations.

Since none of these methods gives a complete picture of the life of a wolf pack, we have created our own imaginary pack. If you like, imagine yourself a scientist or an invisible, odorless observer among the wolves, or even a wolf that belongs to the pack, although you must remember that a pack member will not necessarily be aware of everything in the picture. However you think of yourself, or if you simply read what follows, try to be aware of the different ways people might have acquired the information we present about wolves. What methods might some traveler or researcher or hunter or wolf neighbor have used to find out all these things? And think of the things we mention about wolves that are still puzzles to people. Perhaps you can even think of new ways to find out more about wolves and raise new questions about their behavior.

Imagine a rocky hillside, the top strewn with large boulders. Under these boulders is a long, narrow tunnel just wide enough for a human to crawl through on his or her stomach. The slightly enlarged end of the tunnel is the home of five newborn wolf pups and their mother. The time is spring, and for the next five weeks she will constantly tend and nurse them, barely leaving except for a stretch in the sun when she joins the rest of the pack and a quick trip to a nearby stream for water. Whenever she wiggles her way back through the tunnel and nears the pups, they squirm toward her body for food and warmth.

Although wolves can dig out their own dens, this one was made by a fox several years ago. Unlike that fox, this wolf mother will have the help of her pack in nourishing and teaching her young. They will help her chase away the grizzly bears who come to steal food. They will bring her food, and later when the pups are about five weeks old and weaned, the pack will bring them food too. When the pups are old enough, they will be taught to hunt with the others.

A few days after they are born the pups can hear sounds, and by the end of two weeks they open their eyes. By three weeks they can crawl through the tunnel and play and explore near the den entrance. When they are a month old

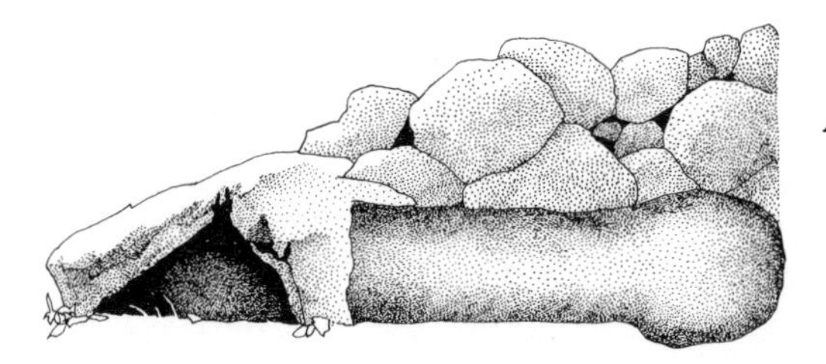

A cross section of a wolf den and tunnel

Wolves greeting each other

they can hold their ears erect, and they begin to look more like wolves. One day the first pup to utter a strange drawn-out howl is so surprised by the sound that it tumbles over backward and runs back into the den. It does not stay inside long, though, for sounds of the rest of the pack returning from a hunt penetrate the dark tunnel, and it tears out into the early-morning light to join the others. The pups are leaping and rubbing their noses against the faces of their father and four other adult wolves, two males and two females. They roll on their backs, yip, squeak, and wag their tails. Some of the grown wolves wearily tolerate their behavior; others act almost as rambunctious as the pups. The wolf mother is given a moose leg and hungrily begins to eat, while the others straggle off to favorite resting spots in the area around the den entrance and on the rocks above. Time to sleep and digest the twelve or so pounds of moose flesh each of them ate earlier that morning.

The pack had hunted and eaten well, and even had enough left over to bury several large fleshy bones a few hundred feet from the den. They are there for the mother, who can't travel with the pack quite yet and needs food to keep strong enough to nurse her pups for another week or so.

They spend this day sleeping and playing with the pups, who always seem ready to wrestle and bite one another or any of the older wolves who will put up with them. As they play, they use many of the lunging and neck-biting movements they will need when they are old enough to hunt with the pack.

Later in the afternoon the pups' mother walks over to her mate, who is dozing in the sun. She rubs her nose along his face, waking him to follow her down the hill. The two of them trot to the river, chase through the shallows after anything that moves, catching nothing, not even trying, just soaking in the pleasure of running and playing with each other. Suddenly their ears twitch at a faraway sound. Long, faint howls too distant to come from their pack pull them from their play. The two of them turn from the river and quickly climb the hill to their den.

The howl

Within minutes their pack is howling too. One of the wolves begins with a long-drawn-out high moan, then another joins in on a different note, until all the adult wolves are howling, each on a different note, changing to a new one whenever they begin a new howl. This is a time of solidarity for the pack, but for all its intensity it is over in a minute. Now there is a flurry of tail wagging and jumping about. The males cluster around the dominant male, acting like little puppies begging food from their father, rubbing their noses along his mouth, while he calmly, patiently, seems to tolerate their behavior. Wolves don't always howl before a hunt or hunt after they howl, but on this late afternoon in spring their song is clearly a getting-ready ritual, for moments later all the adult wolves but the mother begin to quickly move on down the hill on another hunt.

The way this wolf pack hunts tells us a lot about the way they live together. As the wolves set off to hunt, they often form a line headed by the male leader of the pack. He carries his tail high as he trots at a smooth, steady pace across the open land. In winter the snow makes movement far more difficult, and the wolves take turns breaking a path through the snow, the others following in the newly made trench. The wolves' prey moves more quickly than they can when the snow is soft and deeper than the wolves' legs. Sometimes, though, a strong crust forms on top, and because the wolves have such enormous paws they can move swiftly on top of the snow, while their heavy, small-footed prey breaks through the crust and must struggle to make headway.

This afternoon the hunting group has five members—three males and two females. The females are two and three years old, and although smaller than the males, they are much faster than any of them. One of the males is a pup really, part of last year's litter. He is still learning from the other wolves, and though strong and eager, he is not the hunter he will be in a year or so. He has learned to avoid a moose's flying hooves and where to attack a prey animal, but the older animals know more about how an ambush works and how to follow a wounded animal and wait for its death. They know more about which animal to attack and where it might go to escape.

We can't be sure what leads the wolves to their destination but guess it must be a combination of scent, experience, and possibly even sound. Overhead three ravens begin to fly in a high, wide circle. They know from experience that a successful wolf hunt means leftovers for them to feast on.

Whatever has drawn the wolves to this place has not misled them. As the small pack comes to the crest of a long, low hill, it halts to survey the valley below. Next to the stream they can make out the moving shapes of moose. Moving closer, they see that three of them are calves. Experience says, "An easy mark, if it weren't for their mothers." Moose cows are vicious fighters, especially with young calves to protect. Many wolves carry scars from encounters with them: fractured skulls, ribs and legs that heal but become sources of arthritic pain later on.

A moose lifts her head in the direction of the wolves, but they are downwind from her. She returns to her calf as the wolves continue to watch from the hillside.

A signal we can't discern must pass among the wolves, for all five of them suddenly charge down the hill with a plan worked out. No words or diagrams make it clear, but some plan that comes from knowing how a moose behaves and how each member of the wolf pack performs directs them to the next step. They must decide on a victim. As dogs smell the chemicals a fearful human emits, the wolves sense which of these moose it will be possible to kill. They may recognize a lame leg and know it means "slow moving." They might smell sickness on the breath of an old moose and know it will not be able to run fast and long.

The moose is the largest member of the deer family; it weighs about 1,800 pounds

The wolves approach the cluster of moose quickly and relentlessly. The cows chase their calves ahead and turn to face the approaching wolves. Two of the wolves swing around behind the cows and take off after a calf, as the other wolves keep the attention of the cows. While the wolves lunge at them and confuse them, the mother of the calf now being pursued along the stream has momentarily forgotten to protect her young.

The calf is an easy mark for the two wolves. It tires quickly, and one wolf begins to attack its hindquarters while the other grips its neck with her powerful jaws. It soon collapses, and they move off a bit to wait for it to die. In a few minutes the other wolves join them. Not every attack is this successful. Wolves have been observed to try a dozen times before they bring down one moose.

It is interesting to stop for a moment to compare the hunting methods of a wolf pack with human groups that still hunt for their food. The wolf's way of hunting doesn't sound so different from this anthropologist's account of how the Naskapi, a northeastern Canadian hunting people, hunt caribou:

> On snowshoes, the hunters quickly shuffle away from camp carrying their rifles over their shoulders. The Naskapi walk at a fast and steady pace alternating between putting on their snowshoes when moving in deep snow, and removing them and hanging them over their rifle barrels as soon as they reach a hard and icy surface. No words are spoken. Half running, every man takes the wind, weather and every feature of the terrain into account and relates it to the position of the caribou. Suddenly one of the men stops and crouches, whistling low to the other men. He has seen the herd. Without a word the men scatter in different directions. No strategy is verbalized, but each man has made up his mind about the way in which the herd can best be tackled. Seeing the other men choose their directions, he acts accordingly.

There are hunting differences between wolves and men, though. For one thing, wolves rely on a keen sense of smell that humans can only imagine. Many scientists believe that the ears of wolves are so sensitive that it may be sounds rather than smells that lead wolves to their prey. On the other hand, our eyes are much better than wolves', and it is thought that the eyes of people raised as hunters are even more observant than those of the rest of us. But there is another thing that human hunters do that compensates for their comparatively slow pace and their lack of smell. They are able to articulate many questions about the animals they hunt. Where is the animal now? In which direction is it traveling? How fast is it going? Where is it wounded? (Like wolves, many hunters merely injure the prey and then follow it until it is too weak to move farther or fast enough to escape them.) The answers to these questions come from evidence that has to do with the time of day and year, the wind direction, the nature of the land—is it steep? rocky? covered with snow? Is there blood on the rocks and bushes? How much? What is the condition of the feces along the trail? Wolves no doubt consider many of these things too, but unlike humans, they cannot stand around and argue about the evidence and then decide what to do.

Here is how a hunter in Botswana describes why he gave up following a gemsbok (a long-horned antelope that lives in arid parts of Africa) he wanted to kill. He was following the feces dropped by the gemsbok. He believed he could catch up with this gemsbok because he thought the droppings had been made that same morning and therefore the animal was not very far away. "After about twenty minutes the man stopped and said, 'No, it was made last night,' and abandoned the spoor. Asked what made him change his mind, he indicated a single gemsbok footprint with a mouse track on top of it. Since mice are nocturnal, the gemsbok print must have been left during the night." Another hunting animal might have come to the same conclusion—the tracks were so old that the animal was too far away to catch—through its nose instead of by an intellectual process. There is certainly no doubt both human and animal hunters have to learn from older and wiser hunters how to interpret the information they use, and to watch carefully and for a long time how their elders stalk and kill their prey.

Spoor.
Any track or trail left by an animal. It can consist of footprints, hair rubbed off on trees and bushes, feces, or any other sign of its having been in a particular place.

After our five wolves join up around the fallen moose calf, they begin to tear at it. They spend almost the entire evening alternately eating and resting. In some packs the dominant male of the group might be the first to begin, but when this happens he eats alone for only a few moments before the rest join him. There are times when the group might linger around a large carcass, eating and napping for several days, but during this time of the year they have to think about the mother and pups back at the den. They must bring them food, so they carry what they can in their mouths. The pups will get food that is regurgitated or vomited from the wolves' stomachs. Repulsive as it may seem to us, this is how the pups survive until they can join the adults in the hunt. The stomach of an adult wolf can hold enormous amounts of food. This makes wolves' survival easier, for it means they can gorge themselves, and then, if necessary, not have to eat again for two or three days. The food to be regurgitated is not acted on by their digestive juices to any great extent, so the pups get a fairly appetizing meal.

An adult wolf feeding a pup regurgitated food

As the hunting pack heads home, the wolves are not too weary to make some important stops on the way. The forty or so square miles they hunt and live in is *their* territory. For the time being they will not tolerate any other wolf packs or lone wolves on their turf. Their way of putting up NO TRESPASSING signs is to urinate on as many rocks and trees and bushes as they can, giving warning to other passing wolves that a member of the pack that lives and works there will fight to protect its claim to be there. As warm and friendly as the pack members are with one another, they have little tolerance for outsiders and might even kill a strange wolf who trespasses into their territory.

This scent marking probably has other equally important information in it for other pack members who may be passing by or for pups who need some sense of the boundaries of their territory. It could tell members of a partially scattered pack that the others are regrouping for a hunt, or simply give information about who has just passed by, how many, when. Wolves have even been known to urinate on wolf traps, poisoned bait, and human garbage such as tin cans and bottles. Could they be warning other wolves of some danger they have detected? Perhaps today's message read, "Five wolves, recently feasted on moose calf, heading home."

Leaving messages behind is a sign of solidarity with other groups of your own kind. There is a wonderful human example of marking similar to wolf marking provided by the Gypsies who wander about Europe. They have to warn other Gypsies of enemies, inform them of friendly towns, and relay information they have gathered that might be useful in fortune-telling or trickery. According to Jean-Paul Clébert in his book *The Gypsies*:

> . . . mutual aid among the Gypsies requires that, when a tribe has stayed near a village, it should leave a message addressed to other tribes giving the maximum of information that can be used for commercial or provisional purposes. Thus, a Gypsy woman will be the first to go into a farmhouse on the pretext of selling items of linen-drapery, or to tell fortunes. She operates so as to make the owner's wife talk, and she learns about important family matters—the number and ages of the children, recent illnesses and so forth. As she is going away, she will scratch on the wall or mark with chalk or charcoal signs which only her racial brothers will know how to make out. Some time later, when a second Gypsy woman shows up at the farm, she will have every opportunity to tell fortunes and reveal to the utterly amazed and therefore credulous farmer's wife details of their family life.
>
> These signs are obviously very simple hieroglyphs, but in great variety. I have long pledged myself to maintain silence on this subject, but, just as an example, I reproduced *patrin* signs that have already been published in newspapers and magazines. They are not quite accurate but will give a good enough idea of their symbolism.
>
> + here they give nothing
> ≠ beggars badly received
> O generous people

very generous people and friendly to Gypsies
here Gypsies are regarded as thieves
we have already robbed (this place)
you can tell fortunes with cards
the mistress wants a child
she wants no more children
old woman died recently
old man died recently
at loggerheads about an inheritance
master just died
mistress is dead
mistress is dissolute
master likes women
marriage in the air

Gypsy markings serve the same function as wolf markings. They inform groups and individuals of their own kind about other people that are frequently unfriendly.

Returning to our wolf pack on another day soon after the hunt we just described, we find that the mother wolf has decided that the pups are old enough to move on to another home. She has already rejoined the hunting expeditions, but needs to move the pups closer to where the others have been hunting. Soon it will be time to teach them the skills they must have to survive their first winter, but they are not yet strong enough to travel the great distances covered by the grown wolves. They have already begun snapping at flies and stumbling after mice and will soon be able to catch them and other small animals. For now, one of the unmated wolves will continue to stay and watch the pups whenever their mother goes off to hunt.

Although wolf packs have anywhere from three to twenty-five members, it is a rare one that has more than one mated female. This means that each year there is only one litter of pups born in any one pack. In years when food is scarce or the wolf population high, a pack may not produce any pups. Some of the wolves may be older children of the dominant pair. Others may be brothers and sisters of the lead pair, and often a group contains wolves that are not related to any of the others. The relationships among wolves differ from pack to pack.

As there is a top-ranking, dominant male wolf, also called the alpha male, there is a female hierarchy or ranking system as well. The alpha female is usually the mate of the alpha male, who is in most cases the biological father of the pups born each spring. The dominant female has such authority and control over the other females that she prevents them from mating and becoming pregnant by fighting with them to keep them away from the alpha male. The dominant female, however, may mate with other males besides the alpha male.

Outside the wolf pack itself are other wolves, lone wolves, who do not appear to belong to any pack. They may attach themselves to a pack and be tolerated in its territory enough to be allowed to follow, eat the remains of its kills, but not live with the group. These lone wolves sometimes spend their time with other "non-packed" wolves, but still do not establish groups that produce pups. This is not to say that they will never join one or haven't been part of one in the past. They may be young wolves looking for a pack to join. They may be wolves waiting for a territory to open up where they can establish a new group. They may be too old or weak to be good hunters, former alpha animals even, who have been rejected by their old pack. They could be looking for mates. No one knows why some wolves live in packs, some on the edges of them without being fully part of them, and others live alone. We do know that wolf packs are not unchanging collections of wolves that always behave in predictable ways.

The composition of most wolf packs does not stay constant

all year. Once the pups begin to be more self-sufficient, the wolves may split apart for a time, coming together occasionally to hunt. With the approach of the mating season in late winter, the group re-forms and generally stays together until the new litter is almost fully grown. Many of the pups who survive their first winter leave the pack around this time to join a new group or simply go off in search of one.

Even within a litter of wolf pups, observers have noticed that the animals differ in boldness, ferocity, and meekness. Like adult wolves, the pups develop their own ranking order among themselves, although these hierarchies change frequently until the pups are nearly grown. Adult wolves are different from one another too, so that each pack has its own character, depending on the personalities of the wolves in it. The alpha pair may be especially aggressive and not be tolerant of other members and particularly ferocious toward outsiders. Another pack may be led by kinder, more gentle wolves, which give it a much more easygoing, less aggressive character. Especially severe winters and a shortage of food may put pressures on a group that make its members angry and hostile.

Wolf packs are made up of individuals that often have different parents. Since most of the pups probably leave their original pack, each pack is ensured variety of color and size among its members, and even among pups of the same litter. Some are black, others tan, reddish, gray, even spotted.

It is easy to distinguish the five pups in our imaginary wolf pack. The smallest is a reddish-brown male, while the two biggest wolves are a gray male and a black female that resemble the alpha male and female of the pack. There is also one female pup that is gray with black spots and one particularly energetic pup that is also spotted but has a bright white streak on its chest.

If it were possible for one person to follow these five pups over the first two years of their lives, we might find something like the following. It is likely that at least one of them

Wolf pups playing and eating in front of their den

will die of rabies or from an attack by a grizzly bear or an enraged moose. In a particularly lean year one or more of the pups might starve to death. Two or three of the stronger wolves might wander away from the pack and join other packs; one or two might stay with their parents' pack. By the time the pups are two years old, they will not all be together and some will be hunting and living over 100 miles apart. They might never see one another again. And if they did, would they recognize each other? Would any puppy affection remain? Would they recognize and honor their parents? We don't know. Wolf social life is still very much a mystery, even though people have spent thousands of hours studying them.

What emerges, however, is a portrait of a society that is intense, warm, nourishing, and very protective. For an animal whose survival depends on finding large animals to eat, the cooperative hunting techniques developed by the pack are the best and most efficient means of killing prey. Mother wolves and young pups also need the help of adult wolves to bring them food when the pups are too young to move about safely in the open. Wolves need one another.

Quite apart from these reasons of bare survival, wolves have another reason to live in packs, rather than as solitary creatures. People who have watched wolves in the wild are struck by the warmth, playfulness, and loyalty they show one another. Just as we enjoy the companionship of other humans, the pleasure of living in such an intense and affectionate group must be as important to the wolves as the hunting and protective benefits.

LEMURS

A Voyage of Discovery

Sifaka lemurs are one of the most peace-loving of animal groups

The study of animals in their natural environments poses frustrating problems for human observers. We can't change the nature of our senses so that we can see, feel, and smell what the world is like to an animal such as a wolf or a cat or a moth. But even though each animal we might want to observe lives at a pace and on a scale that is not ours, there are many things we can learn about animal life, and in the process learn about human adaptability and ingenuity.

In order to learn about wolves it is necessary to think about how to cope with the vast territories they inhabit. A different set of problems is posed by trying to study the social life of animals that live high up in the trees of tropical forests. One such animal, the sifaka lemur, spends most of its life thirty to fifty feet above the ground and flies through the trees with the grace and skill of a trapeze artist.

Little is known about the life of lemurs. Only rarely do they exist in zoos outside their native Madagascar, and they are now threatened with extinction. It is difficult to get to them and easy to wonder why one should bother to study their lives. What might be interesting to learn about lemurs, worth a trip to Madagascar? One focus comes from the fact that lemurs are like monkeys and are not monkeys, like people and are not people; resemble dogs, bears, cats, and raccoons, and are not any of those either. They exist only in Madagascar and their neighboring Comoro Islands, and developed, as most of the plant and animal life on Madagascar did, without contact with life on the major continents. Madagascar used to be a part of Africa, but about 100 million years ago

Madagascar and the Comoro Islands lie off the southeast coast of Africa

The Lesser Mouse lemur can fit in the palm of your hand

it separated from the continent and became an island, either because of an earthquake or some other major earth movement.

There are many different lemur species. In size lemurs range from the tiny Lesser Mouse Lemur, which is only five inches long and is the world's smallest primate, to the Indris lemur, which is about the size of a wolf and looks like a bear with the snout of a fox.

Lemurs have noticeably large eyes, especially in those species that feed and are active at night, and though they cannot see color, they are extremely sensitive to the slightest variation in the intensity of light. They perceive hundreds of differences in shade and illumination in their environment, where people would only see lights and darks.

Other noticeable features of lemurs are their almost uncannily human hands and feet. They have opposable thumbs, which allow them to hold and manipulate things.

Most lemurs live high up in the trees of the dense rain forests of Madagascar, though there are some species, like the ring-tailed *Lemur catta*, that travel on the ground in small bands of about a dozen animals.

The particular lemur species that will be focused on in this chapter is the sifaka. They are tree-dwelling and must be one of the most peace-loving of animal groups. In fact, the pacifist nature of the sifaka life is what led us to dwell on this particular group of lemurs.

The Indris lemur is almost three feet tall, the largest of the lemur species

Even though lemurs have existed for millions of years, and have been collected by naturalists for hundreds of years, the first detailed study of lemur behavior was done by Alison Jolly in 1963. Since then, there have been a dozen other studies of lemurs in their natural environment, and a lot has been learned about their social life and their relationship with other animal species in their environment. Still, the study of lemurs is new, and there is much more to learn than is already known. For a person who finds the study of life an adventure the lemurs provide a wonderful subject.

Imagine yourself a young natural scientist lucky enough to get a grant to take a scientific voyage to study the lives of lemurs in the tropical forests of southern Madagascar. You have to prepare yourself, to read all the past research and observations about lemurs you can find in order to avoid foolish misinterpretations of their behavior. A good place to start would be with Alison Jolly's books and articles. She is a pioneer in the field and included in her book *Lemur Behavior: A Madagascar Field Study* the daily notes she wrote while observing lemurs in the field. The field notebook is one of the most important tools of the natural scientist and certainly the cheapest. The notebook is used to record and sketch what you observe, and to speculate on answers to the questions that you choose to focus on during your adventure in the field. These notes will provide the basis for drawing conclusions about lemur life when you are back home. You may also see patterns emerge from things you noted down but thought were unimportant at the time.

In another book, *A World Like Our Own: Man and Nature in Madagascar*, Alison Jolly published her conclusions about lemur life and wrote about the citizens of the People's Republic of Malagasy. Her work will give you a sense of the efforts they are making to build a new nation, overcome poverty, and preserve animal life at the same time.

In addition to reading as much as you can about lemurs, it would be useful to look at pictures, to borrow films taken by other researchers in the field, and to visit natural history museums, where you might come upon lemur skeletons, pelts, or even stuffed lemurs. It is unlikely that you will be able to see a live lemur before your voyage, so all the knowledge you take to Madagascar with you will be secondhand. You will not be able to experience lemurs until you arrive at your final destination.

There are other preparations to make. A good pair of binoculars is essential, as is a tape recorder that records sounds at a distance and reproduces them accurately. In a jungle there are many different noises that occur simultaneously, and you will need equipment that will let you pick out the sounds you want to study. Your task will be like trying to record the voice of one person in a noisy restaurant from thirty feet away. You will also want a camera with a high-power lens, and if you can afford it, a Super 8 or 16mm motion-picture camera. And be sure to stock up on film, tapes, and batteries, as there are no shops in the forest.

All this equipment has to be packaged so that you can carry it and so that weather, insects, or an accidental bump or fall will not ruin it. Details like how you pack the equipment can make the difference between a successful or a disastrous expedition. Likewise, forgetting notebooks and pens and pencils will make it impossible for you to record observations, sketch what passed by too quickly to photograph, and record speculations on what you've observed. And forgetting to protect the paper in a waterproof case can lead to an indecipherable soggy mess that you may discover only after arriving home.

You also have to choose clothes, and a sleeping bag and tent; make sure to take a knife and maybe a gun, some rope, and any other things that you might take on a camping trip into the wilderness. You also have to worry about food and heat. Ironically, the simple task of preparing to observe animal life forces people to explore their own needs and to be creative about meeting them.

Finally, before you go you'll have to contact the government of the People's Republic of Malagasy, who will be your hosts, and get permission from them to study in their country. Also, if you're wise, you'll learn as much about your host country as possible before you leave. Whenever you travel to any place to study, you have to develop a warm, respectful relationship with your hosts. A traveling scientist is a guest and has to be considerate of the people who live in the place he or she is permitted to work in.

Unfortunately, your trip to Madagascar will last for only two months, and that limits what you can see and learn. Still, some aspects of lemur life can be studied in two months, such as: Do lemurs recognize one another as individuals? Do they relate to neighbors of their own and other species? Do they get along with one another? Does each band have a leader, and does every member of the band have a rank from top to bottom like chickens and some groups of people?

You have to find questions that interest you and at the same time be prepared for the unexpected. The death of an animal, an unseasonable storm or drought, the invasion of a territory by a strange group of animals, or the tumbling of a favorite roosting tree can change social behavior, and sometimes these serendipitous happenings, though preventing you from learning what you wanted to know, can lead to other unexpected and important discoveries about animal behavior.

The voyage will take you from the United States across Africa to the island of Madagascar. After a brief stop in Tananarive, the capital of the Malagasy Republic, you meet Dr. Rachel Mangabe, a fictional version of many actual Malagasy scientists. She will accompany you on your trip, and will drive you south to the Berenty Reserve. It is one of the few places in the world where the social life of the sifaka lemurs is still intact. Dr. Mangabe has spent years studying lemur behavior, and she works with the government, trying to solve the complex problem of developing a new and poor nation while trying to preserve its unique animal and plant life.

The landscape within the Berenty Reserve is like nothing you've ever encountered, even in your imagination. It is so dark on the ground that unless you climb a tree you can hardly tell the difference between day and night. The ground is covered with spiny bushes and thick brush. All around are insects and flowers you've never seen before. Even the trees aren't familiar. You remember reading when you were planning your trip that 95 percent of the animals and plants on the island of Madagascar are found nowhere else in the world, but you had no idea of how foreign, even unearthly, it would feel to be surrounded by so many unfamiliar living things.

As you step carefully through the brush trying to keep up with Dr. Mangabe and the park rangers who are helping you carry your equipment, you hear cries, see shadows flitting through the shrubs, and hear sounds of animals crashing through the trees. Some animal comes into view—a giant rat? No, a miniature monkey, or maybe a cat. It moves too fast for you to tell, but you remember its large, shining eyes. Another face appears, grotesque and alien, as you hear ghostly sounds over your head. And you recognize somehow that these faces and sounds represent the animals you have come to study.

The Romans believed that the spirits of the dead sought

out the light at night and looked on the living with glowing eyes while they lamented plaintively. The Latin name for these spirits is *Lemures.* When animal researchers in Madagascar encountered strange animals that could walk upright like people, had glowing eyes, and shrieked out late into the night, they named them "lemurs" after these Roman spirits.

The first night in a strange place is always difficult. It took all day to find a campsite, clear it, set up small tents, make sure all the equipment was stored safely, set up a rendezvous with the rangers in three weeks for more supplies and food, thank them as they disappeared through the brush. Alone with your co-worker you feel for the first time how isolated you are from a world that is familiar and comfortable. As tired as you are, sleeping that first night in the middle of an animal reserve in southwestern Madagascar is almost impossible. Everything seems alive, threatening. It is as if the animals you have come to watch are observing you and trying to learn whether you are dangerous or friendly. At last you fall asleep, only to awaken the next morning to discover what seems like a miracle: several animals fifty feet above you sitting on a tree branch looking as if they are praying to the sun. It is no miracle, however, just a troop of *Propithecus verreauxi verreauxi*, the sifaka lemurs you have come to study.

A sifaka lemur sunning itself

Dr. Mangabe, who, since you have spent time together, insists upon being called Rachel, explains that the sifakas sun themselves every morning upon rising. They sit in sunning spots in the trees, legs spread, with arms outstretched in praying position. When the sun warms their sparsely furred bellies, their underarms and thighs, they slowly turn around and sun their backs. You are not the only one to find this sunning like a religious experience. Many Malagasy communities tell legends about sunning sifaka lemurs. They claim that the lemurs are incarnations of the spirits of their ancestors, who, though dead, continue to worship the life-giving sun in animal form.

The sifakas do not sun for long. Once warmed up, they

begin to groom themselves. They lick and scrape their fur, using their teeth to pick out insects and clean it.

From the ground the sifakas resemble monkeys, but when you look closer it is clear that they are not monkeys. Their thighs are long and muscular, more like human thighs than ape or monkey thighs. Their snouts resemble those of dogs or bears; their long tails are fluffed a bit, like cats' tails. Their hands have long, thin fingers that could be those of a tall, lean person, and the second toe on each hind foot has a claw that sticks up at a right angle to the foot. Lemurs use this claw (all their other toes have toenails) to scratch themselves on the shoulder or behind their ears, just the way dogs scratch themselves. When you first see lemurs, it's easy to get the impression that they are unreal creatures pasted together from parts of many different animals, just like imaginary beasts or monsters.

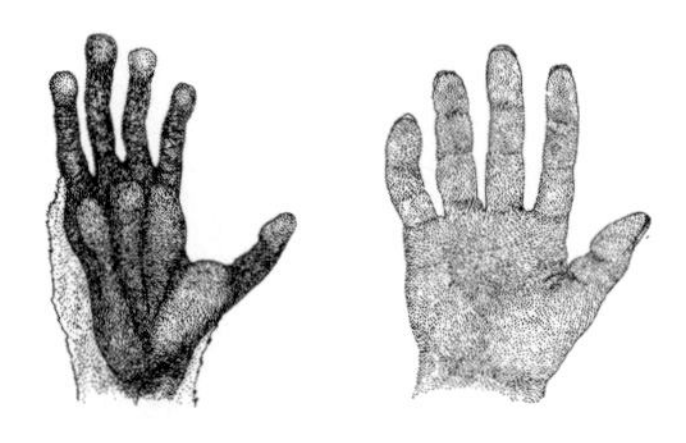

The hand of a sifaka lemur is similar to a human's

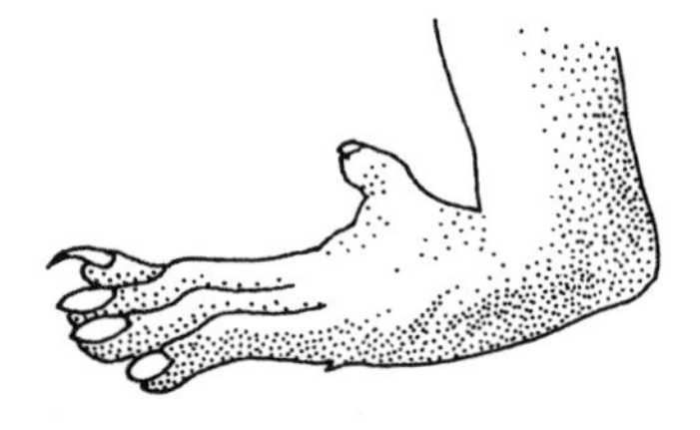

A sifaka lemur's hind foot with grooming claw

The general impression of a sifaka group is one of a family on a picnic. There are usually four to nine members in a group, which might consist of two adult males, two adult females, a baby, and a young male and female. You can't tell from the ground how many lemurs are in the group you saw sunning. It's not easy to look up into the tropical sun or to pick out all the animals in the foliage. You have to invest time to find out about these animals, which spend 95 percent of their time in trees and are on the ground only on the average of 72 minutes of the 1,440 in each day. And you have to begin recording your impressions and observations.

Here are excerpts from an imaginary field journal, one which you might keep during your voyage. Of course, every individual scientist's field notes differ. Writing field notes is a bit like writing a diary. You have to pick a form and style that is comfortable and casual. You can include impressions, can abbreviate things, and don't have to worry about spelling, grammar, punctuation, complete sentences, or other constraints of formal writing. You can address questions to yourself, raise issues to think about when you return home, and make suggestions for future observations. However, there is one major difference between a personal diary and scientific field notes. You have to report your observations as carefully and accurately as possible, sometimes in great detail, and with sketches and diagrams. Even if what you see puzzles you or contradicts what you expected to find, you have to report your observations exactly and as objectively as possible. Loyalty to the specifics of observation instead of to a favorite theory is an essential characteristic of being a scientist.

The fictional diary presented here, the one you might have written, is based on Alison Jolly's field journals of 1963 and 1964, and her more recent observations in the Berenty Reserve published in *A World Like Our Own*. It is also based on other reports of lemur life and other field journals we read while doing research on this book. It is the kind of journal you would keep on your voyage and the kind you can experiment with if you care to study ant, dog, cat, people, or other animal behavior in your own community.

FIELD NOTES. EXCERPT NO. 1: *After first full day in the Berenty Reserve*

Spent most of today unpacking equipment, checking out the camera, tape recorder, organizing our camp. We are under trees over sixty feet tall, in a little clearing, like a pocket in the forest floor. Occasionally we catch some sun, but usually it's dark down here. Rachel has been orienting me. We spent several hours walking away from camp in different directions. I had to find my way back. It was lucky she was with me. I haven't learned to navigate by the sun or mark a trail. Wish I'd done more camping when I was younger.

Just before sunset we took a tour of sifaka territories. Rachel has a map that points out the territories of the different sifaka bands. It was first sketched by Alison Jolly in the sixties and has been added to and confirmed since then.

Sifaka bands have pretty fixed territories that aren't very big—about 1/10 of a kilometer squared. That's about 315 feet by 315 feet, or the area of two football fields. We passed through four territories, and Rachel said each of them contains a band of from six to ten sifaka lemurs. The territories are covered with giant kily trees, though there are lemur trails that run through the brush and shrubs under the trees. I think I got a glimpse of several sifakas, but I still have trouble picking them out.

We walked the boundary of the four territories, and Rachel picked out the crossing points where sifaka bands make forays into other bands' territories. These crossing points are at the meeting point of high branches of the kily trees, so a group can move from one territory to another without having to touch the ground. Rachel said the crossing points usually have a clear view of ground trails and paths through the trees. They seem to have been picked out to give invading groups good lookout points to warn them if the band whose territory they're invading is around.

Rachel also says that the invasions are never too serious and that if I'm lucky I'll see the sifaka ritual before my stay is over.

She drew me this picture of sifaka band boundaries that have remained the same since the sixties, when they were first studied:

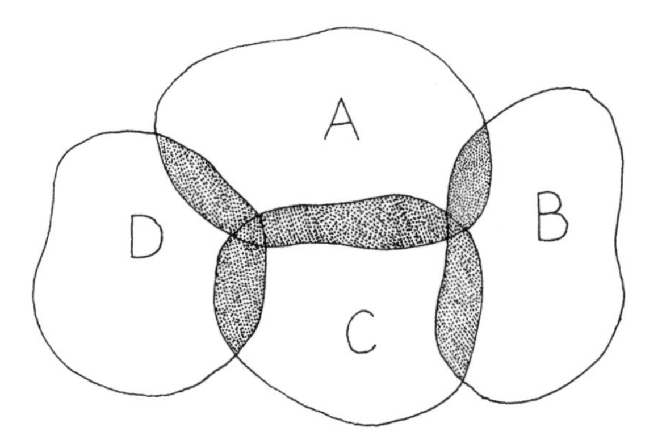

The white areas are never disputed by the sifakas. Each band has its exclusive use of these areas. The shaded space around the borders shows the limits of the movement of one band into another band's territory. All this is still abstract to me. I hope to get to know one territory well and to experience the relationship between lemur bands.

It's been a long day. The next few days we'll look for the sifakas in the territory that I'll be studying—territory D, where our camp is.

FIELD NOTES. EXCERPT NO. 2: *After three days of observation*

Been here in the reserve for three days now and can finally pick out my lemurs. There are six of them. As best as I can make out, there are two older males, one looking very scruffy with chewed-up ears. I haven't seen any fighting yet in the group, but something must have fought him. There are also two older females, but I can't tell the difference between them, and can't even tell the sex of the two younger lemurs.

The troop doesn't stay still for long, and I can't move well in the brush.

At 11:00 this morning I lost the troop. At 1:00 I surprised it very close to the ground, and they began screaming at me as if I were a predator. The sound was like *si-fa-ka si-fa-ka*, and Rachel told me the sifakas got their name from this warning scream. They climbed a little higher up the tree and started screaming louder, moving around in the trees, mobbing me the way blue jays mob a cat. They were trying to chase me off. I know they won't harm me and that I'm supposed to be an active observer, but they scared me to death and I wondered for a minute if I wasn't crazy for coming here. Rachel also told me that when she was a child some of the older people used to tell the children in her village that the sifakas knew who was bad and that when they screamed they were giving a warning to be good.

So far I just have a general impression of the group. Only Old Chewed-up Ears is familiar to me. I always look for him first and try to see who's around him. Here are a couple of sketches of how the animals related to one another during the times I could see them today. By the way, I think I'll name the male with chewed-up ears Scruffy.

I can't find any definite arrangement within the group. All the males and females seem to treat one another equally. If there is any dominance it seems that the females lead the way more often than the males, but I can't be sure. There seems to be no inner group fighting. It's too early to tell anything definite about their social order. In this the sifakas clearly differ from many monkey troops and from other small social groups like wolves.

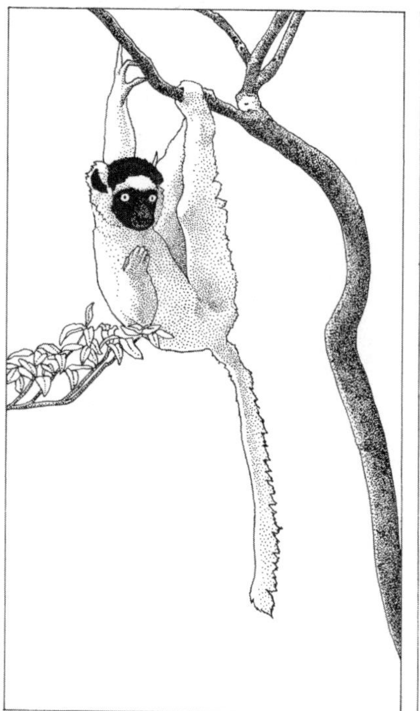

Sifakas eat fruit, leaves, and flowers

Sifakas sleep high in trees

FIELD NOTES. EXCERPT NO. 3: *After five days of observation*

It finally happened this afternoon: the sifaka ballet. I've seen films of the sifakas leaping from tree to tree, but the real thing is practically indescribable. Scruffy was staring at me directly over his muzzle. Then without warning he leaped backward off the tree. He seemed to double in size as he did a full somersault in the air and landed clinging to the trunk of a tree thirty feet away. He looked calm, as if he had never moved. A leap of thirty feet, sixty feet high in the air with no net, no sense of effort—I wondered what a circus would think of a troop of lemurs.

Questions to think about for tomorrow: What does Scruffy do all day? How to recognize other members of the group? Look again for signs of leadership, and for contact with other groups. How can I find ways to increase my territory to follow the group?

FIELD NOTES. EXCERPT NO. 4: *After seven days of observing*

I noticed several hawks high in the sky this morning and prepared my tape recorder and directional mike. The hawks are the sifakas' only natural enemy—other than man of course. I wanted to tape my group's response to hawks and wasn't disappointed. As one hawk descended, the group let out a series of roars and screams—completely different from the noise they made when they mobbed me the other day. It was so loud it hurt my ears. Got it all on tape, and now maybe I can get my band to mob me again. I'd love to have the two series of sounds to compare when I get home.

Thinking of home—Rachel leaves tomorrow and I'll be on my own. She says the lemurs have accepted me as a harmless human and not a hunter so there is nothing to worry about. I don't know. . . .

FIELD NOTES. EXCERPT NO. 5: *Eight days of observing, first day alone in the reserve*

Been very careful about marking trails, finding my way around. Lemurs aren't on my mind so much as being alone. I'm not much of a scientist today—all I can think of is wanting a hamburger and fries—played a tape of the Rolling Stones today. Snuck it along in my baggage, but was afraid to take it out when Rachel was here.

FIELD NOTES. EXCERPT NO. 6: *Nine days of observing*

Back to my lemurs today. Can easily pick them out. They don't seem to mind me, can't even get them to mob me. Made a trip into territory A to get mobbed so I could record it. It took several hours, but I got the sounds I wanted. A good comparison with the hawk sounds, only how can I be sure that the mobbing sounds of the group I taped are the same as my group's? Maybe a trip to a third troop's territory will give me another mobbing tape. If it isn't too much different from what I got today, I'll feel confident about my comparisons.

A ring-tailed lemur

FIELD NOTES. EXCERPT NO. 7: *After fifteen days observing*

I'm feeling confident moving from territory to territory, and I can now find three sifaka bands with ease, though I still can't distinguish all the members. Got a delightful unexpected surprise today. Was watching my band with binoculars when I heard a crashing through the brush. I expected a lion or tiger or herd of elephants, even though I knew intellectually that these animals don't exist in Madagascar. Guess I've never been able to conquer my jungle fears. Anyway, I turned around in time to see a dozen *Lemur catta*, their ringed tails raised high in the air, moving through the brush. My sifakas looked down at them, but didn't mob or make too many sounds at all. They must have recognized the *Lemur catta* as non-enemies. These two different lemur species seem to be able to share the same environment without conflict—or is it the same environment? *Lemur catta* have the ground, the sifakas have the treetops. It seems to be a convenient spatial arrangement.

FIELD NOTES. EXCERPT NO. 8: *Three weeks of observing*

I decided to take a day off from my observations—I know the troop now—the easiest way to distinguish members is by the shape and color of the fur above their brows. The variations among lemur faces is so great that they probably recognize each other as individuals. (How could I find out if this is actually true?) One way to distinguish males and females is that the males have a scent gland on their throats, and they often can be seen throat-marking trees, defining their presence through the trees and high bushes.

The variety of facial markings make it easy to identify individual sifakas

The cap line and the cap stand out as easy ways to recognize individual sifakas.

There is a lot of speculation about the function of the lemur's cap. One theory I read about is that it is not just a patch of fur but actually a sense organ—a heat sensor that works like the black pigment of a lizard's third eye or a heat-sensing missile or a thermostat. I've seen sifakas change their behavior according to the temperature—praying in the morning sun, crouching when it gets hotter so their white backs reflect the sun, and seeking shade in the noon sun. How is it possible to discover whether the cap has something to do with sifakas' heat sense? This is something to think through when I get home—need to find out more about heat sensitivity in other animals, about thermostats. Maybe I can design some experiment that would reveal more about the sifaka cap—it might even get me another grant to come back to this wonderful place . . . Anyway, back to the caps.

Scruffy has a deep brown cap with a symmetrical cap line. The other older male has a jagged, irregular cap line and a red-orange cap. He's Carrot-top to me. One female has a chocolate-brown cap; the other has a very light caramel-brown cap, so they've become Brownie and Candy.

I discovered an easier way to distinguish the youngsters. One has a twisted leg. It must have been broken in a fall. It's very playful and active; Vulcan seems right. And Sparky will do fine for the other youngster.

So there's my group: Scruffy, Carrot-top, Brownie, Candy, Vulcan, and Sparky. It's funny how naming creatures can make them much more familiar. I see individuals now, not just a group.

Questions for further observation and speculation: Does the group always stay together? How are the animals in the group related to one another? Are the young ones children of the older males and females? Is this an extended family group, a group that forms just to forage together (I haven't seen the sifakas eat anything but plants, and what is more astonishing to me, they seem to drink no water at all. Maybe dew on the leaves and liquid in berries is all they need)? Why do the animals stay together? What could indicate this? Maybe watch for signs of playlike activity that has nothing to do with sex, food, or violence.

One more note. Taking this day off was wonderful—something I must do at least once a week.

FIELD NOTES. EXCERPT NO. 9: *After four weeks*

I've been watching my group's territorial border for the last three days. The first day was uneventful—not really, it gave me a chance to look at flowers and insects that grow nowhere else in the world. But no lemurs—maybe a glimpse in the treetops, but I still can't be sure. I did hear cries that I've come to recognize as those of the sifakas, but nothing else.

The second day of waiting I picked out a foraging group of lemurs coming toward the crossing point. It must have been the group that occupies territory A. They passed overhead, scent-marking all the trees, the males with their scent glands and urine, the females by rubbing their genitals against the trees. They waited until they were into my group's territory before stopping to eat, rest, and groom. Sifaka grooming seems to be a major social event. The sifakas clean one another, pick out insects, lick and comb one another's hair. It reminds me of the way my older sister and her friends used to do up one another's hair for hours at a time. Here, however, the males groom males and females, the young groom the old, the old groom the young, and females groom males. The rarest type of grooming by my count is female-female grooming. Perhaps this goes along with the dominant role females seem to play in the group. This needs to be studied more. What is the role of grooming in keeping the social group together? Sometimes it seems as if grooming is a form of greeting, like shaking hands or bowing. Would grooming have this social role of communicating the idea "I don't want to fight you" as well as having a cleaning function? Most likely, as grooming seems to play a pacifying role in monkey, cat, and some other mammalian groups as well.

I decided to do something crazy as soon as the troop was settled. Maybe I'm beginning to feel like a lemur, but all of a sudden I wanted to climb a tree and sniff the scent the males put on the trees with their scent glands. I did it too, only as soon as I got thirty feet above the ground I lost my bearings, couldn't find the spot where the marking was, and everything smelled the same to me. When I got back to the ground, I started laughing at the thought of telling my parents that I was thirty feet up in a tree sniffing around for lemur smells. It seemed hysterically funny.

I realized tonight that I'm lonely—it's hard being out here by myself, and wonderful, too. Tomorrow I'll go up the tree again, find a scent-marked place, and take back a sample too. Maybe, if I can't sniff it, at least there might be chemical traces I can analyze back home.

FIELD NOTES. EXCERPT NO. 10: *After four weeks, one day*

Got it. Smelled it. Have six samples of scent-marked bark. Don't know what to make of it, but it gives me some material to work on when I return. Maybe Rachel and her colleagues have already done work on scent-marking. I'll ask her when she returns to pick me up.

FIELD NOTES. EXCERPT NO. 11: *After four weeks, three days*

Today I witnessed full warfare. Carrot-top spotted an invading group of lemurs. (Maybe "invading" is the wrong word, since I don't know enough about borders—I have to be careful about using words like "invading" that assume more than I've actually observed.) There were five of them. It happened so fast I only noted that one had an unusual yellow cap and another was melanotic. I'd heard about melanotic sifakas, and they're as beautiful as Rachel claims, with their dark-brown sideburns and chest and brown patterns on their forearms and thighs that look like the wings of a great hawk or eagle. I don't remember much about the others, the noise was so great. It was a *sifaking* noise, but seemed much higher and louder. I couldn't really tell, as I've never heard eleven lemurs screaming before. Even if it gets to be a nuisance sometimes, I'm glad I force myself to carry my tape recorder all the time.

Melanotic.
Having unusually dark-colored hair or skin.

The males scent-marked wildly, on one branch, then another. The females also marked with urine and feces. The two groups faced each other, and it felt as if a battle was about to begin. What I witnessed was closer to a chess game. The two troops of lemurs faced each other from opposite stands of trees and leapt toward each other in close fashion. They continued scent-marking; then, instead of engaging each other and fighting, they seemed to leap about at random. They didn't fight each other but seemed to struggle over positions on the trees. The goal seemed to be to occupy spaces facing out from your own direction. The only physical contact I observed was even more unusual. As both troops seemed to have reached an agreeable boundary, the melanotic male, jumping backward, landed on Scruffy's back, grabbing it with hands and feet. I watched for biting and hitting, but an amazing thing happened. The attacker, instead of biting Scruffy, made several upward scrapes of his head in a grooming gesture, and they disengaged without any hostile physical contact at all. Then both lemurs retreated to their own group. It was as if the hopping, screaming, and scent-marking were lemur forms of some delicate diplomatic negotiation, with all parties fully aware that they didn't intend war.

FIELD NOTES. EXCERPT NO. 12: *After five weeks*

This past week, play has been the subject of my observations. Since I've been here, there have been no instances of physical fighting in Scruffy's group or between groups. Nor is there any top animal in the troop, though the females seem to take the initiative in finding food, claiming sleeping branches, and other activities. There are even times when neighboring groups sleep in trees next to each other, and one gets the distinct impression that all the lemurs in a territory know one another as individuals. Before she left, Rachel told me that some observations indicate that sifaka groups are much more informal than I imagined, that they exist mostly for foraging, raising the young, and/or companionship and play. They also seem to interbreed and to break down and regroup with new members.

Sifaka lives are peaceful except at breeding time, when the males fight viciously. Their only enemies seem to be the hawks I saw during my first week here.

People, of course, are also another enemy of the sifakas, and lemurs are powerless to do anything about guns or about the destruction of the forests for human settlement or the cultivation of crops. Human friends, I suppose, will have to find some way to preserve some bands to ensure the sifakas' survival. Rachel is obviously such a friend. She has been working to ensure that some sifakas will survive into the next century, just as they have survived over millions of years.

But enough of this—I'm supposed to be writing about my observations of the sifakas, not pleading for their survival. Play is what I've been watching, and it's important to be accurate about what I saw before I leave next week. Sifaka play is like a pantomime show. The animals hang down from branches by their tails and try to box or wrist-wrestle—only each animal uses both hands and feet. There are eight limbs to watch in this playful game. It all takes place in slow motion, the opposite of a cat fight, where blows come fast and are meant to hurt. This is pure play and goes on until one sifaka gets tired and leaves. Two specific instances of play I observed today:

—At about 10:00 this morning Candy approached Vulcan, who was sitting on a branch. She made a cuffing gesture, but it was slow with no force behind it. He responded by trying to grab her wrist. They both dropped to hang from the branch by their feet and spar with their hands. This lasted about seven and a half minutes and led to their grooming each other.

—Scruffy and Carrot-top began to wrestle while they were grooming each other. They grabbed at each other's wrists, nose, and ankles, trying to get through each other's guard. However, the effect was different, say, than baboons or dogs, which go at one another furiously. It was all done slowly and with soft gestures that deliberately avoided being painful. There was no winner or loser, nor was there in any other sifaka play I observed.

FIELD NOTES. EXCERPT NO. 13: *After six weeks*

Rachel returned. My stay is about over, just as I've become comfortable with my band and with being alone. We talked—no, I talked and showed Rachel my notes. We agreed to spend a week before I return home at the University of Tananarive sharing our perceptions of lemur life and considering what research needs to be done. Rachel also suggested that we take a long walk to the edge of the reserve tomorrow. She said it was beyond lemur territory and promised a surprise.

FIELD NOTES. EXCERPT NO. 14: *After six weeks, two days*

Rachel doesn't make idle promises. We spent most of a day walking through the brush. Sometime in the afternoon I sensed a rank, terrible odor, not a scent-marking odor, much stronger. Then there were squeals, shrieks, sounds that didn't exist in any of the sifaka territories I studied. Rachel pointed toward a tree that seemed alive. It was crawling with thousands of indescribable creatures. She clapped her hands, and all the creatures took off—thousands of bats with wingspans of three, four, even five feet, flying through the air. Because of their size they are called flying foxes, though their scientific name is *Pteropus rufus*. I don't like bats, never could study them, but these were different. The bodies were red like foxes', their wings were as wide as eagles' wings. And

A Pteropus rufus bat

there were trees full of them. No wonder lemurs never wandered into bat territory.

At night Rachel and I talked about bat and lemur territory, about the Berenty Reserve. She said that as far as scientists can tell, bats, lemurs, birds, trees, plants—all the life in the reserve has reached a balance. Somehow the space of each is respected, and all the populations are stable and healthy. This complex equilibrium must be studied more—it could have something to teach us about our unstable world.

FIELD NOTES. EXCERPT NO. 15: *On the plane returning home*

What to make of what I've seen and experienced? I miss Rachel and wish I knew more about her life. But it's lemurs I came to study and ought to write about. As a group the sifakas seem to challenge many assumptions about what keeps animal or human society together. My observations are clearly limited. There is just so much that can be learned in two months. Still, the nonviolent nature of sifaka life is striking. They seem to have evolved a small, cohesive group life based on play, grooming, and foraging. Neighboring groups play-fight, and individuals seem to recognize and greet each other either through scent-marking or rubbing noses. Sex does not play a major role in their lives except for a few weeks which I didn't observe. During those weeks people report serious struggles over females, the development of dominance relations, and frequent injuries—all those aspects of behavior that seem to characterize social groups bound together by sex and power relationships. Perhaps the size of the group makes the difference, or the ease with which they can obtain food in the trees, or maybe it's their relative freedom from predators other than hawks and people. Whatever the key notions are, I must be careful not to overgeneralize from my limited observations.

It is difficult to avoid the conclusion that one of the main bonds of sifaka social life is mutual affection. For that reason alone (and there are many others), it is worth continuing the study of sifaka life in its natural setting, as well as protecting the sifakas from extinction.

The imaginary journey ends, like most scientific voyages of exploration, with a few answers and many new questions. How do sifaka groups live over a period of years? How do they deal with death or injury of a member? There is much to be learned about mutual affection in answering those questions, as well as considering others, like what happens to the group in drought years or in times of fire or pestilence? Are there any occasions when different bands of sifakas join together for some common tasks? The variety of colors and shapes of sifaka caps and cap lines indicate substantial interbreeding of groups. When, where, and how often does this happen? And do offspring grow up in the band of their parents? Do children recognize their biological parents and have any special regard for them? All these questions are currently being studied in the reserves and in several zoos in the Malagasy Republic.

As these field notes indicate, observing the social life of animals takes time and patience. You have to look for signs of social interaction among animals within a group and contact between groups. Greetings, grooming, play, fighting, all give hints about the nature of group life. A picture of the whole social structure of group life emerges only after piecing together years of observation. Imagine, for example, the false conclusions about sifaka life that would be drawn if scientists concentrated on the mating season or, equally, if they never observed it.

Observations of social interaction are not always as easy to obtain as they are for lemur groups. It is more difficult, for example, to observe elephants or termites, the former because of their size and the range of their territory, and the latter because of their numbers. Imagine what it would be like to record all the social interactions in a city with a population in the millions of termites after spending time in the comfortable company of six or eight lemurs. In the next chapter we'll try to make that transition and explore a termite city that numbers in the millions, a mini-New York, London, or Tokyo.

TERMITE CITIES

The scale of termites' lives is completely different from our own

There are termites living in the redwood forests near our house. The Pacific Dampwood termites are pale, whitish insects, almost half an inch long, and live in crumbly, dead wood. Whenever we expose their nests to the light by picking them up or turning them over, the termites squirm and wiggle deeper into the log. On examining the wood carefully, we can see the tunnels and hollow chambers they have made. Because these termites' main food is wood, they actually eat their rooms and passageways into existence. Inside their intestines, microscopic protozoa break down the wood so that it is digestible. When termites clustered in the same spot continue to eat, the room gets larger and larger. Eating becomes building. When they need more food they find more wood—and build more rooms and connecting tunnels.

Microscopic protozoa. One-celled animals, invisible except under magnification, that live inside wood-eating termites and digest the wood, making it usable by the termites' bodies.

With lemurs it is easy to feel that it is only the barrier of language that prevents us from understanding them. If only it were possible to talk with them, we could know them intimately. But it is hard to even conceive of communicating with one termite, much less a million of them.

Our lemur investigation was difficult at first, but soon it was possible to tell one lemur from another. They were also aware of *us* as individuals, and eventually, as they learned to trust us, they ignored us, so that we felt we were watching their daily lives as they would live them whether we were there or not.

With termites it's different. We're too large to get inside a log with them or follow their movements within their nests. The scale of their lives is completely different from ours. Yet termites, though small and perhaps not particularly interesting as individuals, have, like other social insects, a complicated community life. People have been interested in the curious ways these insect societies seem to resemble our own, and are awed that so simple a creature could have developed such a complicated way of life.

Because most species of termites are sensitive to light and dry heat, they hide deep within their nests and are extremely difficult subjects to study. Investigators have had to break into these nests and disrupt the life inside in order to learn about their societies.

Two views of a homemade termite cage

There are other, less disruptive ways to learn about the relatively simple colonies of wood-eating termites that are found in North America. One is to seek out a colony in an old log and put the entire chunk of wood in a large glass container and watch it day by day. People have kept such colonies for years. A better arrangement would be to sandwich sheets of paper and wood between pieces of glass, introduce a termite colony to it, and observe what they do. If all goes well, they should live fairly normal lives, and you won't even have to find food for them since the nest wood and paper are their food too. This can work better than a log because the nest can be easily pulled apart and examined. Still, we can't become termite size and get inside the nest with them, or follow them through the course of an ordinary day.

There is, though, one important part of the life cycle of all termite colonies that is possible to watch without disturbing the insects at all. Once or twice each year large numbers of winged male and female termites leave their nests in order to mate and begin new colonies. For many termites the event takes place on days when the air is warm and still and the sun bright. Others, especially those in Africa, wait for the first autumn rains. The termite soldiers make a small opening in the nest and crawl out into the daylight to patrol the area

Winged male and female termites leaving their nest to mate

around the hole. They stay there, ready to attack anything that threatens their nest and the thousands of brothers and sisters still deep in the ground. In the passageway below the soldiers, there are cluster after cluster of winged male and female termites nervously running and colliding with one another until suddenly hundreds of them pour out from the hole and fly up into the sky.

Throughout the afternoon, thousands of termites burst forth like this. They have spent their entire lives seeking safety in darkness, but for a few brief moments the light draws them out as they fly as far up and away from their homes as they possibly can.

As soon as each termite lands, it begins to beat its wings up and down so vigorously that as the ends hit the ground the base of each wing is torn from the body and falls away.

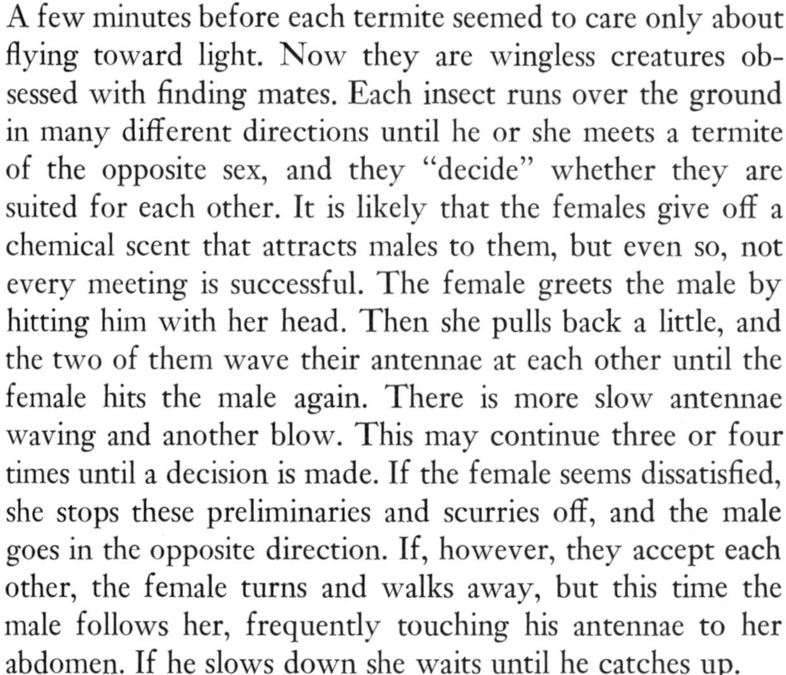
A few minutes before each termite seemed to care only about flying toward light. Now they are wingless creatures obsessed with finding mates. Each insect runs over the ground in many different directions until he or she meets a termite of the opposite sex, and they "decide" whether they are suited for each other. It is likely that the females give off a chemical scent that attracts males to them, but even so, not every meeting is successful. The female greets the male by hitting him with her head. Then she pulls back a little, and the two of them wave their antennae at each other until the female hits the male again. There is more slow antennae waving and another blow. This may continue three or four times until a decision is made. If the female seems dissatisfied, she stops these preliminaries and scurries off, and the male goes in the opposite direction. If, however, they accept each other, the female turns and walks away, but this time the male follows her, frequently touching his antennae to her abdomen. If he slows down she waits until he catches up.

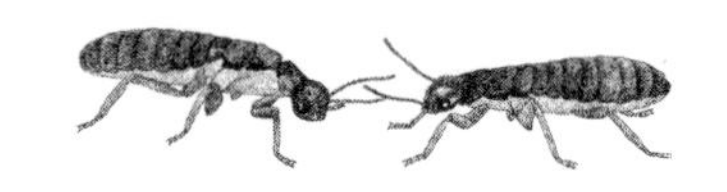

Macrotermes termites waving their antennae

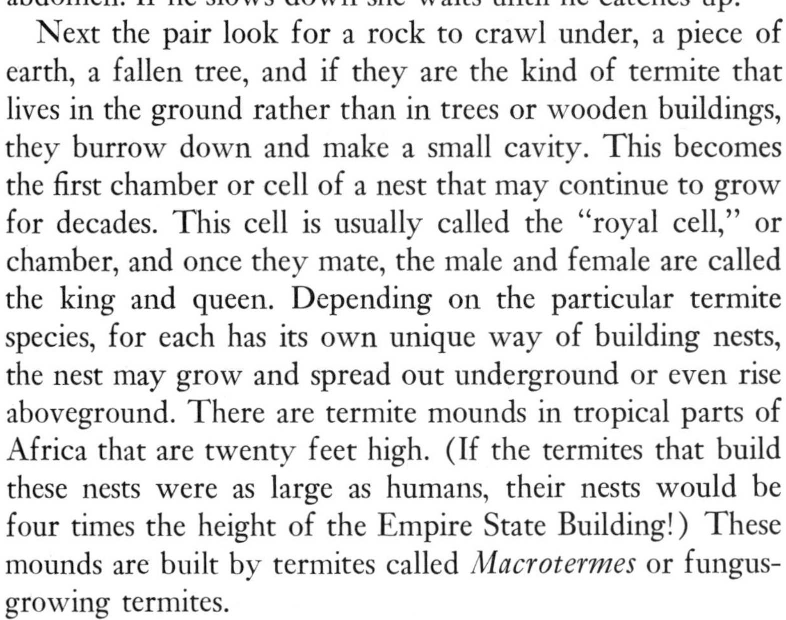
Next the pair look for a rock to crawl under, a piece of earth, a fallen tree, and if they are the kind of termite that lives in the ground rather than in trees or wooden buildings, they burrow down and make a small cavity. This becomes the first chamber or cell of a nest that may continue to grow for decades. This cell is usually called the "royal cell," or chamber, and once they mate, the male and female are called the king and queen. Depending on the particular termite species, for each has its own unique way of building nests, the nest may grow and spread out underground or even rise aboveground. There are termite mounds in tropical parts of Africa that are twenty feet high. (If the termites that build these nests were as large as humans, their nests would be four times the height of the Empire State Building!) These mounds are built by termites called *Macrotermes* or fungus-growing termites.

Of course there are no twenty-foot-high termite mounds near our house, nor are there any on the whole North

American continent. To experience and learn more about complex termite societies that build these nests it is necessary to travel to Africa, Asia, or Australia. There it is possible to encounter landscapes dominated by termite cities.

Imagine you are part of a group of natural scientists taking a trip around the world, not to experience restaurants and human creations, but to be in the presence of animals in their natural environments. You have visited Indonesia and watched 400-pound tortoises emerge from the Pacific Ocean to lay their eggs in the sand and then retreat back to the sea. You have been guided through the Serengeti Preserve in Tanzania and seen elephant herds and prides of lions, and you have come to Lake Manyara National Park in Tanzania to see the mounds made by the *Macrotermes* builders. They seem like enormous lumpy castles squeezed out of wet sand. As you move close to one of these mounds, there is little that indicates there is life inside. The group has already decided to do anything possible in order to find out what goes on in this nest, even though everyone was warned that the soldier termites can bite. Fortunately, the expedition brought along heavy gloves, several long sticks, a flashlight, a magnifying glass, and a shovel. And, of course, notebooks and pencils and an insect guidebook to help identify the different termite castes.

In order to prepare the group to observe a *Macrotermes* city, the guide has sketched a history of how termite colonies grow, as well as provided information on the termite caste system, much in the way a tour guide would give the history of a cathedral or a medieval castle.

The mounds you are observing, like all termite colonies, grew from a single royal chamber. The termite queen lays her eggs in this cell that she and her mate have dug. The larvae that hatch from these first eggs are workers. Unlike newly hatched ants and bees, they are not legless and com-

Caste.
Among social insects, a group that has a distinct body type and performs specialized tasks within the colony.

Larva.
The first form of an insect after it hatches. Larvae are usually wingless and often wormlike and have little resemblance to the adult insects.

pletely helpless. While they need some help at first, they can move around by themselves. They look like small, slightly different versions of full-grown termites. As they mature they go through several molts. They shed their skins and emerge each time larger and slightly changed. Their parents, the king and queen, feed them regurgitated food—leaves, seeds, sometimes animal material—but after one or two molts they are able to find their own food and feed themselves. Termites, like most other social insects, continue to feed one another all their lives. They even have a way of asking for food by stroking fellow termites with their antennae.

Many of the highly organized *Macrotermes* species raise some of their own food in fungus gardens. Worker termites leave the nest to find leaf and wood particles which they mix with feces to make compost heaps in special chambers in the nest. They plant spore filaments of mushrooms on the compost which then grow into mushrooms. The larvae are particularly fond of these mushrooms, and they make frequent trips to the fungus chambers to eat.

The worker caste contains males and females which will never become sexually adult because their sexual organs never mature. They spend their days and nights hollowing out more nest chambers and coating the walls with a substance made from their feces. They also take care of their newly hatched brothers and sisters. They pick up the eggs as they are laid and carry them off to special chambers in the nest.

When there are enough workers to do all the work of the colony, the king and queen stop caring for young and building the nest. They stay in one cell and produce fertilized eggs. For the rest of their lives they are fed and cared for by their worker children. In many species the queen cannot even get out of this royal cell because her abdomen grows so large she can no longer fit through the entryways. In fact, the queens of the species *Macrotermes natalensis* have abdomens that increase to about 14 centimeters long. Eventually this kind of termite queen is able to deposit 30,000 eggs a day,

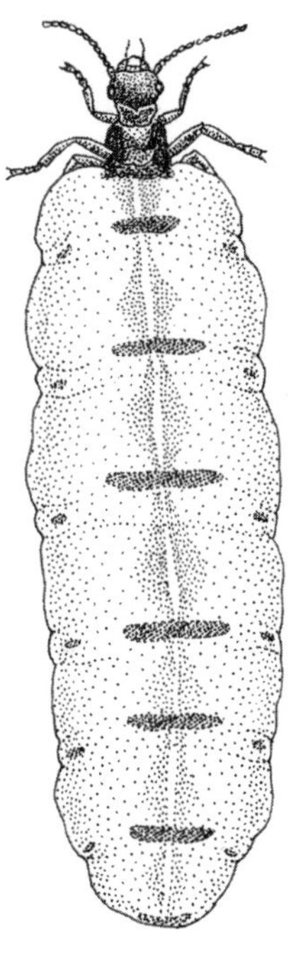

A macrotermes queen can lay 30,000 eggs a day and can live up to twenty years

which comes to almost 11 million eggs a year. Since the queen of this particular species lives for at least ten years, she might lay 110 million eggs in her lifetime.

The workers of some termite species also venture outside the nest at night to hunt for food and bring it back to the nest. To give a sense of the magnitude of work that goes on in a termite colony your tour guide told of a man named Edouard Bugnion, who once watched a five-hour termite food march in Ceylon. The workers walked about twenty yards from their nest to a group of trees, where they collected lichen from the trees and humus from the earth. They spent the entire night doing this, not returning to their nest until after sunrise. Bugnion estimated that 1,000 termites walking ten abreast passed a particular spot every minute, a total of at least 300,000 insects.

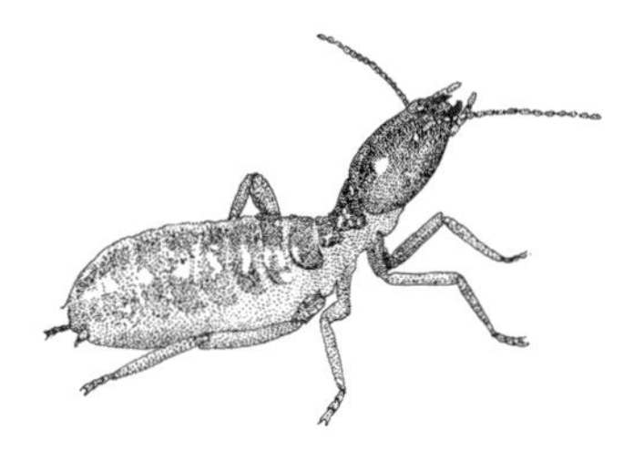

A macrotermes worker

The worker termites make up the largest group or caste within a termite nest, but there are three other kinds of termites: soldiers, which protect the nest; winged males and females, called reproductives; and a small number of male and female termites called supplementary reproductives. Supplementary reproductives are capable of turning into kings or queens whenever a king or queen dies.

After a year or two, the colony begins to produce a few soldiers, or warriors. These termites vary from species to species, but they all have very special ways of defending their nests. They usually have extremely strong mandibles or jaws which they use to bite intruders. In some species the soldiers' heads have large, thick coverings, and many species emit either poisonous liquids or sticky substances that cover and kill small insect enemies. In fact, their heads are usually so overdeveloped that they can't feed themselves and they must rely on workers to put regurgitated food into their mouths.

Although all the larvae of the different castes look alike when they first hatch, the ones that will eventually become sexually mature are called nymphs. These are the termites

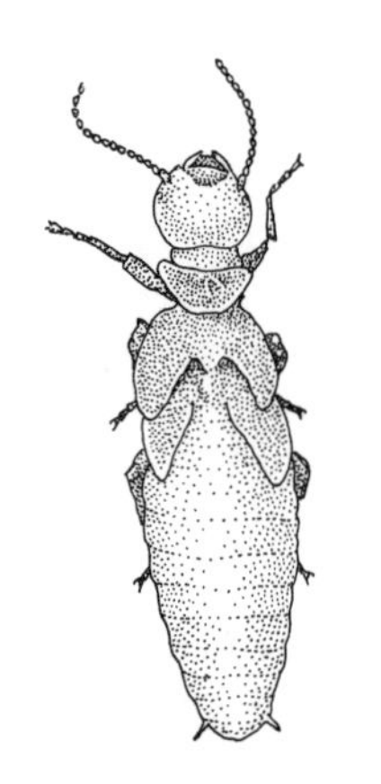

Macrotermes supplementary reproductives can turn into kings or queens

that either develop wings and leave the nest on a "wedding flight" as their parents did (reproductives) or turn into supplementary reproductives. The supplementary reproductives usually appear only when a king or queen dies, and they do not exist among all termite species. They probably live in another form until they are needed. When something happens to the king or queen, the chemical atmosphere in the nest changes, and this change seems to cause a young nymph to develop into a new king or queen. It is also thought that they may develop even when the king and queen are alive and healthy, but that they are killed by workers or the king if they are not needed. In this way, the colony may continue for centuries, provided it is not destroyed by a virus or fungus disease or outside predators.

As the number of termites in the colony of a complex termite species such as the *Macrotermes* increases, so does the size of the nest. In a couple of years a mound begins to appear above the level of the ground; eventually it climbs higher and higher. Unlike anthills made of loose dirt that can be squashed by feet or washed away in a rainstorm, these termite nests are covered with a hard, cementlike substance. In ten years or so, a *Macrotermes* nest may have 10 million insects living inside.

When you know this much about termites, it is natural to want to see all the castes functioning, to observe the queen's chamber and the fungus gardens, to see soldiers and workers, as well as the winged termites waiting for their nuptial flight and the supplementary reproductives waiting for the death of a king or queen. However, confronting a termite mound is quite different from looking for lemurs in the trees or seeking out wolves in the tundra; you know where the termites are but can't observe them without disturbing them.

After admiring the hard, crusty surface of the mound you've chosen to study, it's time to get bold and poke a tiny hole in it. With a sharp, forceful jab the sharpened end of a stick pierces the nest shell. Almost immediately there is feverish activity—heads appear and scurrying legs disappear and reappear again. A huge-jawed soldier runs out of the hole and clambers around the outside edges of it. Soon he is joined by another. Although your group is only a foot away, the termites seem to have no sense that you are near, for they can neither see you nor hear your voices. As one of your group pushes the stick back near the hole, it is immediately attacked by the soldier termites—they bite the stick, clinging to it ferociously, as they try in vain to destroy it. Even chimpanzees have learned the way termites respond to sticks. During termite mating seasons when there is an escape hole in the nest, chimpanzees find long sticks (sometimes even stripping them of leaves and twigs) and work them into the escape tunnels. A chimp waits a few moments for the soldiers to bite the stick and then withdraws it and eats the clustering soldiers so quickly that they don't have time to attack the chimp's mouth.

A macrotermes soldier with powerful jaws

In this part of Africa humans also enjoy a termite snack now and then. (Roasted termites are said to be tastier than shrimp.) The people there have two ways to catch the insects. The first is to jab a pole into the nest much as the chimpanzees do. The second method involves trickery. The *Macrotermes* of East Africa begin their mating exodus soon after the first rains of autumn. People have discovered they can imitate the sound of the rain by beating on the outside of a nest with sticks. In Zaire the termite hunters make a slapping noise with their tongues to imitate the sound of rain. If all the other conditions of temperature, time of day, and season are right, the termites are tricked into responding as if it were raining, and as they leave their nests they are immediately captured.

Once the wall of the mound has been broken open, there's something for your group to observe as workers begin to appear inside the hole. It is possible to make out an abdomen here, an antenna there, but what they are doing is not altogether clear. Eventually, though, the opening grows smaller,

A young chimpanzee learns how to use a tool to get termites

and moving closer with a magnifying glass, one can actually see a worker deposit a drop of excrement on the edge of the hole. Shortly after, another worker places a piece of soil on the drop of excrement. Most of the work, however, is done from the inside, so we see only parts of bodies passing inside the opening and the slow narrowing of the hole until at last it disappears.

No one knows exactly how termite work is organized and how they know what to do. How is an alarm given in a termite mound? How do soldiers recognize enemies? How do workers know where to go to repair the damage? These questions require more observation and perhaps the study of chemical and even electrical signals on such a small scale that we don't yet have fine-enough instruments to detect the signals that function in the community. We do know, however, that termites are extremely sensitive to air currents and that when a hole is made in their nest, they are immediately alerted by the air hitting their antennae. Whenever social insects—bees, ants, wasps, or termites—are alarmed, they usually emit a chemical that has a particular smell that acts as a warning signal to other nestmates. In the case of termites, whenever the nest is damaged, those termites near the hole probably begin to emit this chemical from their abdomens as they run from the hole into the nest. When they run into other termites, they return with them to the hole, or the new recruits follow the trail. This journey back and forth is repeated many times until there are enough workers to repair the damage.

Termites have one other alarm signal that may be used to transmit danger signals over greater distances. Although the workers and soldiers of most species are blind and deaf to noise that travels through the air, they are extremely sensitive to vibrations that travel through the ground. Sometimes when there is danger, termites will bang their heads very quickly against the sides of the passageways and chambers of their nests, sending warnings throughout the colony. Many years ago it was discovered that buildings near railroad tracks or noisy factories seldom had any termites in them. The reason is very likely that there are too many confusing signals from the machines to allow the termites to live normal termite lives. There are too many danger signals.

So far your group has caught glimpses of workers and soldiers and observed them carrying on their normal functions. However, the inner workings of the nest have been inaccessible. Part of the group's initial plan for observation was to take apart a large portion of the nest layer by layer to get a picture of the entire termite society, but after watching the defense and repair job, some people begin to feel a little closer to these creatures and are reluctant to disturb them so drastically. As in most groups of natural scientists, there are a number of internal disagreements about responsibility to the creatures being studied. One member of the group feels that destroying a ten-year-old city built entirely by such small, hardworking insects which weren't harming anyone is simply not fair. A second person is afraid that you might be attacked by hordes of soldiers and that you should not push your luck any further. Another one thinks that you will disturb the life of the colony so much by what you plan to do that you wouldn't really learn anything about termites' lives anyhow. It would be no different from excavating an ancient palace and making guesses about the daily lives of people who lived thousands of years ago. The last argument put forth is that the mound should be dissected. Termites are small, mindless creatures. There are millions more in colonies all over the plain, and you really don't know if the ones you would disturb will be injured or not. They might be able to build a whole new nest after you leave. And what difference does it make, anyway? People shouldn't let feelings get in the way of knowledge.

After considerable debate it is agreed to continue the investigation a little further by making a larger cut into the nest and seeing how much more it is possible to observe about termite society. The city, however, will not be destroyed.

Since the upper, towerlike extensions of termite mounds are part of the nest ventilation system, the insects rarely live in these parts, except while they are building or repairing them. To create minimal damage and disruption, it makes sense to choose a section below this and make a narrow slice from top to bottom so you can see inside the main mound.

As soon as the mound is opened, it becomes possible to see hundreds of worker termites in individual cells. They are frantically wiggling and beginning to form clumps at the end of every cell as they form lines leading from one cell to another. A few soldiers appear to be moving along the edges of the exposed cells, antennae scanning, probably looking for the cause of the destruction. Clearly, people are so big that the termites are not aware of them. There's no anteater's tongue, no aardvark's claws for them to attack.

The light has by now sent the last of the workers into undamaged cells, where they can no longer be seen. The cells will soon be sealed off from the outside world. The real life of the termites is again closed to us. Your group is left to examine the abandoned cells of the fragment cut from this termite mound. Since the cells are lined with hard coatings, they keep their shape without crumbling much, and you can have a good look at them.

One part of the fragment is coated with fungus. You have stumbled upon a termite farm. It is also possible to make out cells and narrow galleries leading to and from the cells and farm. The hole has the appearance of a miniature abandoned city, and one can imagine it teeming with termite life. It is just a small part of a mound but contains more termites than New York City contains people.

It is time to head back to town. In some ways the group really hasn't seen much—no swollen queen laying twenty eggs a minute, no chambers of winged males and females, not even a worker carrying an egg from the royal chamber to a nursery cell. But it was worth the trip just to see this incredible landscape totally dominated by mounds raised by millions and millions of tiny creatures.

It is likely that if you had actually taken such a voyage you would have even more questions about termites after seeing the enormity of their world and appreciating the organization and work that went into building these mounds than you would simply by reading about termites. What are the differences between how we plan and build our houses and cities and termites theirs? How do they communicate with one another? How can there be so many different types of termites—workers, soldiers, etc.—all from eggs laid by one queen? Many of the questions we had about lemurs also come to mind: What holds the group together? Do the termites recognize one another as individuals? Are there any termites who are loners? Does the group benefit the individual termite, or does the individual exist for the sake of the survival of the group?

It is possible to find answers to most of these questions, but it would be impossible through on-site observation by one person or one group of natural scientists. As in the case of wolves and lemurs, it takes the patient work of many people over a long period of time to piece together a view of the social life of an insect community. People have learned what is currently known about termites through many different methods of observation, both in the field and in the laboratory. Moreover, there is much we have yet to learn and possibly much we humans will never know.

There are four kinds of social insects: bees, wasps, ants, and termites. All insects that live in colonies have three things in common: there are different castes or types of colony members that are physically different from one another, and only one caste produces the young—all the rest are sterile; there is more than one generation of insects in the colony; and the members of the colony cooperate in taking care of the young.

Bees, ants, and wasps are now known to be related to one another in the sense that they evolved from a common ancestor. Termites have the same ancestor as another ancient insect, the cockroach, and are not related to bees, ants, or wasps. In fact, one of the most interesting things about studying social insects is that two separately evolved branches of the insect family have independently developed similar ways of living in highly structured groups. This provides an interesting contrast to the lemurs, which evolved independently of monkeys and apes and developed *different* social forms.

Although social insects build complicated nests and can communicate with one another, they are not what we consider intelligent creatures. They have memories and learning ability, but they can't use what they know to solve new problems. They have adapted to living together in ways that have over millions of years become more and more complex. Fossil evidence suggests that insect societies have evolved from solitary insects, which may or may not have cared for the larvae, to the complex ones that we know today, where several generations and physically different castes live together. Just how they manage to build nests the way they do and care for the eggs and larvae and collect food is still not clear.

For the past 350 years people have suspected that social insects communicate through chemical messages and receptors. Many scientists today believe that as soon as we can analyze this system we will understand most of what there is to know about the behavior of these insects. It is only very recently that anyone has begun to test this theory.

Receptors.
Nerve endings that are sensitive to particular stimuli or messages. (Ears have nerves specialized to receive sound waves, noses are sensitive to odors.)

A number of scientists have begun to collect and chemically identify the different substances that come out of termites' bodies and act as messages. Many of these messages are emitted by the queen and to some extent by the king. The messengers are all the other inhabitants of the colony. Here is how communication between the royal pair and the rest of the colony takes place: As the workers move in and out of the royal cell to pick up eggs to take to the nursery cells, they stop to take care of the queen. Part of the care consists of licking her body. Researchers have tested this behavior by taking a small amount of the queen's hormone and rubbing it over a piece of wood. When the wood is put near worker termites, they immediately begin to lick it, as if it were their queen, and continue until all traces of the hormone are gone.

Hormone.
A substance that is produced by an organ in an animal and moves to another part of the body, where it causes something to happen because of its chemical activity. In some cases, as with termites, the hormone leaves the body and affects other members of the colony.

What makes a termite do this is still a mystery, but it is possible that a worker termite licks the queen because she tastes good. Whatever the motivation, the result is that the worker now has in its stomach a small amount of whatever hormones or chemicals cover the queen's body. When any worker meets a fellow worker which places its antennae in such a way that the worker-with-message regurgitates food from its stomach into the mouth of the other termite, part of that message from the queen is now transmitted into the stomach of the second worker. In this way messages can eventually spread throughout the entire colony. We do not know how each termite is able to interpret these messages, but can guess that among other things, the hormones act to stimulate or retard the growth and development of each individual that receives a portion of them.

Several years ago another scientist wanted to find out about how different kinds of termites are attracted to one another. He placed workers, larvae, and soldiers in separate glass tubes and covered each opening with a piece of gauze. Then he placed different kinds of termites on the area at the top of the tubes to see if there were any tubes they preferred to cluster around. Judging by the crowds of termites that gathered around the larvae, the scientist decided that they must be the most attractive group. The workers in particular were strongly attracted to the larvae. Hardly any termites were drawn to the soldiers. The conclusion we might draw from this experiment is that each group has a particular odor and that perhaps the reason the larvae are given such good care is that their odor, once recognized, is very pleasing to workers and draws the workers to them.

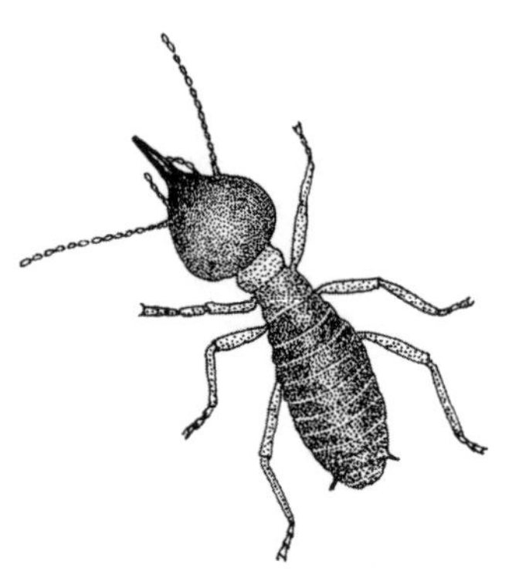

A Nasutitermes exitiosis soldier

Termites not only respond to chemical indicators and scents, but use them as weapons as well. One species, *Nasutitermes exitiosis*, has warriors with long, armored foreheads. When they attack an enemy, they shoot out a gluey liquid that coats the intruder—an ant or an anteater's nose, perhaps. An ant cannot get the glue off and becomes so entangled it eventually dies. Unfortunately for the termite, this glue frequently splatters back on the warrior too, with the same result.

Another termite species has a soldier with an elongated

upper lip that can emit a strong toxic chemical. This soldier brushes the enemy with his lip and covers it with the poison. Although termites are also affected by the poisons of different termite species, they do not appear to suffer from the ones they or their own warriors emit.

One of the aspects of the social life of termites that is still not understood has to do with the caste system. Termite larvae and nymphs go through several molts before they turn into soldiers, workers, reproductives or supplementary reproductives. It appears now that, at least in the lower termite orders, the king and queen termites emit a substance that prevents the sexually undeveloped pseudergates from becoming fully developed. This hormone travels quickly through the colony as termite feeds termite.

Pseudergate.
Among the lower termites, a special kind of worker termite that has either molted backwards and lost its wing buds or developed from a larva. Pseudergates can molt further and develop into supplemental reproductives.

Most colonies need only two productive termites. What probably happens eventually is that when either the king or queen dies and stops emitting this chemical, new reproductives develop to take their places. But we still don't know what makes some larvae turn into soldiers and others into workers. There is even evidence that some nymphs who have begun to sprout wing nubs and hence to become sexual forms occasionally molt backward and lose whatever features they had begun to acquire. Are the larvae fed differently by the workers? Is there some genetic difference among the eggs? From what is known right now, most people think the strongest influences are the handling of the eggs and larvae by the workers, combined with the chemical atmosphere and temperature of the nest.

Termites clearly recognize the different castes in the colony. Smooth functioning depends on soldiers protecting the others and not attacking them, but particularly their protecting the royal pair. Workers recognize eggs and larvae as things to care for. We have no way of knowing yet whether termites are capable of recognizing other nestmates as individuals. Many scientists who study animals think that an important difference between social insects and birds or mammals who live in groups or societies is that only birds and mammals have the ability to recognize individual members of their groups. It is certainly possible that this is true. Since termite colonies are so large, in most cases their lives so brief, and they are genetically as alike as brothers and sisters, it may be that termites have no awareness of individuals. This lack of awareness might be one reason their colonies function as well as they do. But at the same time, it is hard to forget the rituals of termite courtship and the fact that there is some individual selection. There could be some aspect of termite behavior that our senses and our scientific measurements have not been able to pick up that does allow for individual recognition even in the mass world of the termite.

CONCLUSION

The World of Social Animals

We have made our presence known to the creatures we share the world with

There is no place on earth where wolves, lemurs, and termites share the same environment. Nor are there any people who have spent a significant amount of time studying all three. We have spread ourselves across the earth and in one way or another have made our presence known to almost all the creatures we share the world with. However, our understanding of the living world is limited by our size, by the powers of our senses and the instruments we invent to aid them, and by the disruption of life created by any attempt to study it.

It is probably no exaggeration to say that wolves have learned as much as they need to know about us. Most people are dangerous to wolves, and they often respond to people as enemies. Moreover, they range over a territory where people cannot follow on foot. Airborne surveillance of wolf life is possible, but it is disruptive too. Wolves are aware of unusual presences just as people are, and their behavior is affected by the fact of being observed.

Lemurs are also aware of people who try to study them. They have the advantage of height and mobility which gives them the power to decide how much of their life will be revealed to people. Termites are only dimly aware of human observers. Because of their size and the fact that they live underground in darkness, the closer we try to observe their everyday lives, the more we disrupt the behavior we want to learn about.

The three animal social worlds we looked at, and many more social worlds, are interesting to study for themselves.

The pleasure of understanding them is probably not that different from the pleasure one can get from understanding a complex piece of music, a painting, or a poem. And there is another component to that understanding that may be equally important. By studying something outside of yourself, you can also get insight into your own life. The social life of animals provides us with models of how life is organized, models that may be positive or negative but that we can use to measure our own social lives against.

We, as people, are blessed and cursed by culture and choice, by the ability to imagine how we would like to live and the occasional will to make our dreams real. We are born into a particular culture, into a particular world created for us by other people. As we grow we can accept and affirm the social world we inherit. Or we can look about us, study other worlds both animal and human, and build possible worlds that seem better or fairer or more pleasant than the one we were born in. The ability to conceive and plan a future, as well as to observe and learn from other forms of life, is a particularly human phenomenon. It makes the study of different forms an important human venture. We study wolves, lemurs, termites, and other species for the pleasure of knowing about them but also for what we can learn about enriching our own ways of living.

For all the limitations of our size and senses, we do learn things about animal life. We know that the three groups we looked at in this book differ in very important ways.

Sifaka lemurs live in small bands with no great differences in the status of group members. The leaders of the group differ from situation to situation, though females tend to be the most dominant individuals. Their social world is very democratic compared to the structured world of the wolf pack with its clearly defined leaders.

Termite society is even more structured than wolf society. It is a caste society with each individual having a biologically determined role within the whole group. We don't know, for example, whether individual soldier or worker termites differ or not, but the general impression of termite society is that it is made up of type-cast individuals that, within each caste, do not differ from one another.

Each of these social groupings served its members well, at least before the intervention of humans. Wolves roamed most of the Northern Hemisphere and were able to hunt and survive in the Arctic and as far south as Mexico. Lemurs lived in Madagascar's tropical forests in harmony with their environment, neither overpopulating nor destroying their or other animals' food sources. Termite colonies have spread throughout the world.

With the growth of human population, animal worlds have been disrupted. It is ironic that as people began to learn more about animal life, they also began to destroy that life. The growth of cities and the cultivation of large plots of land combined with hunting and trapping have reduced the wolf population and may destroy it.

The lumbering and burning of the forests in Madagascar to produce farmland and timber, which is much needed in the People's Republic of Malagasy, have driven many lemurs from their traditional habitats and destroyed many other animals as well.

The threat of development to lemur life is an instance of the dilemma of human growth within a world shared with other living creatures. The people of Malagasy are poor; the nation is underdeveloped. The replacement of forests by farmland and the displacement of lemurs are not comparable to the sport hunting of wolves. The former is being done out of a need to feed people and provide them with adequate housing, while the latter is unnecessary and serves no need of human society. There are also attempts to develop lemur preserves, such as the Berenty forests, so that lemurs do not become extinct, as they easily could. Yet it is painful to contemplate the cruel trade-off that balances the survival and growth of one form of life against another.

Even termites are being threatened. Researchers at the U.S. Forestry Service's laboratory in Gulfport, Mississippi, have been able to create methoprene, the hormone that helps determine the caste of a termite larva when it molts. They have experimented with giving termites extremely large doses of methoprene. These doses have created bizarre, non-functional termite mutants, such as creatures with soldier's heads and worker's bodies. These mutants disrupt the flow of information and food in a colony, and within several weeks it is destroyed.

The hormone also kills the organisms that live in termites' digestive systems and are essential to the digestion of wood. Thus, mass starvation is another consequence of an overdose of methoprene.

The use of methoprene to destroy termite colonies might be beneficial to some people whose homes are threatened by termite damage. However, there have been instances where the use of pesticides such as this one has led to unnecessary destruction of insect, plant, and animal life, as well as to human harm.

Lemurs, wolves, termites, and many other living forms may disappear from the earth due to human carelessness and greed. This can become a lonely planet, one on which even people could not survive. We must respect the creatures we share the earth with and never reach the point of enjoying hurting any living thing for the pleasure of it. Life feeds on life, and destroying one species can lead to the destruction of an entire environment. The vegetarian sifakas and the termites that live on dead wood do little harm to other living beings, but even the wolf who hunts and kills to survive does not kill for pleasure or in order to overeat. People, though, kill for pleasure, or to acquire more than they can possibly use. We must learn to live modestly and leave others in peace if they are not threatening our lives. We must become part of a shared world, give animals their place while striving to meet our own needs. Then we will be able to enjoy the beauty and diversity of life instead of being overwhelmed by the sorrow and devastation caused by the notion that the earth belongs only to people.

Whenever we consider the study of animals in the wild, we always have to be aware of the effect that we, the outsiders, have upon the lives of the animals we observe. This is simply a caution we must take, not an argument against observation. Because we can relate to the creatures we study, it is possible to learn more about ourselves and to develop bonds with other living creatures that can lead us to a more respectful attitude toward life in general. It is possible to love a wolf pack you have come to know well, to know and care about a lemur band, and to feel respect and awe in the presence of termite cities. There is a romance to being a natural scientist studying animals in the wild, one tempered by a need to be precise and accurate, but one that nevertheless makes journeys to Arctic tundra, tropical jungles, and barren deserts fulfilling.

SOURCES

PAGE

20 William Little; H. W. Fowler; J. Coulson, *The Shorter Oxford English Dictionary on Historical Principles*, revised and edited by C. T. Onions (Oxford: Clarendon Press, 1959), p. 1847.

25 Konrad Lorenz, *The Year of the Greylag Goose* (New York and London: Harcourt Brace Jovanovich, 1979), p. 9.

26 P. P. G. Bateson and Peter H. Klopfer, eds., *Perspectives in Ethology*, vol. 2 (New York: Plenum Press, 1976), pp. 221–2.

28 Edward O. Wilson, *The Insect Societies* (Cambridge: Belknap Press of Harvard University Press, 1971), pp. 421, 447.

36 Farley Mowat, *Never Cry Wolf* (Boston: Atlantic Monthly Press, Little Brown, 1963), pp. 70–2, 78–81.

41 Barry Holstun Lopez, *Of Wolves and Men* (New York: Charles Scribners Sons, 1979), p. 88.

47 George Henriksen, *Hunters in the Barrens: The Naskapi on the Edge of the White Man's World* (St. John's: Memorial University of Newfoundland, 1973), cited in Lopez, p. 88.

48 Richard B. Lee and Irven DeVore, eds., *Kalahari Hunter-Gatherers: Studies of the Kung San and Their Neighbors* (Cambridge: Harvard University Press, 1976), p. 342.

50 Jean-Paul Clébert, *The Gypsies* (Baltimore: Penguin Books, 1969), pp. 244–5.

61 Alison Jolly, *Lemur Behavior: A Madagascar Field Study* (Chicago: University of Chicago Press, 1967).

62, 67 Jolly, *A World Like Our Own: Man and Nature in Madagascar* (New Haven: Yale University Press, 1980).

FURTHER READING

Darwin, Charles. *The Illustrated Origin of Species*, Abridged and Introduced by Richard E. Leakey. New York: Hill and Wang, 1979. Darwin's book is probably the most important natural-history book ever written. A magnificently illustrated version that also includes information discovered since Darwin's book first appeared in 1859.

Fox, Michael W. *The Soul of the Wolf*. Boston: Little, Brown and Company, 1980. An interesting book about wolves and how wolves and humans relate to each other. Many photographs.

George, Jean Craighead. *Julie of the Wolves*. New York: Harper & Row, 1972. A moving and exciting novel about a young girl who spends some time with a family of wolves.

Jolly, Alison. *A World Like Our Own*. New Haven: Yale University Press, 1980. A beautifully photographed account of a trip, sponsored by the World Wildlife Fund, made by a lemur specialist. The book deals with the problems of preserving native wildlife in a country where there is a growing need for agricultural land.

Lopez, Barry Holstun. *Of Wolves and Men*. New York: Charles Scribner's Sons, 1978. The natural history of wolves and how people throughout history and around the world have thought and written and created myths about them. Beautiful photographs and old drawings.

Natural History. A monthly magazine published by the American Museum of Natural History. Regular articles on animals and animal behavior, plants, astronomy, and archaeology. Illustrated with wonderful photographs. It is available from Natural History Membership Services, Box 4300, Bergenfield, New Jersey 07621.

von Frisch, Karl. *Animal Architecture.* New York: Harcourt Brace Jovanovich, 1974. A fascinating and easy-to-read account of animals who build structures, from shellfish and snails to social insects, birds, and beavers. Written by the world's greatest bee expert.